FUTURE SATELLITE GRAVIMETRY AND EARTH DYNAMICS

T0134631

Future Satellite Gravimetry and Earth Dynamics

Edited by

JAKOB FLURY and **REINER RUMMEL**
Institute for Astronomical and Physical Geodesy, Muenchen, Germany

A C.I.P catalogue record for this book is available from the library of Congress

Published by Springer,
P.O. Box 17, 3300 AA, Dordrecht, The Netherlands

www.springeronline.com

Printed on acid-free paper

ISBN 978-1-4419-2131-4   e-ISBN 978-0-387-33185-0

Table of Contents

Earth, Moon, and Planets (2005) 94: 1
DOI 10.1007/s11038-005-2814-5

PREFACE

In 2003, Astrium Space Industries carried out for ESA a study on the future needs for dedicated satellite gravity missions. (ESA contract no 396 2/01/NL/ GS). The title of this study was *"Enabling Observation Techniques for Future Solid Earth Missions"*. The technological part of the study was based on a thorough assessment of the future needs in Earth sciences for more precise and refined gravity models.

For this purpose a workshop was organized from 30 January to 1 February, 2003 at the International Space Science Institute (ISSI) in Bern/Switzerland where a group of Earth scientists exchanged their ideas with the objective given above. This collection of articles is based on the outcomes of this workshop. All articles underwent a review process according to general scientific standards.

The support of ESA and of ISSI is gratefully acknowledged. We also thank Springer's Astronomy and Astrophysics department for the pleasant coopcration during the preparation of this volume.

Munich, 14 June 2005

Jakob Flury and Reiner Rummel
Guest Editors

Earth, Moon, and Planets (2005) 94: 3–11
DOI 10.1007/s11038-005-3755-8

GEOID AND GRAVITY IN EARTH SCIENCES – AN OVERVIEW

R. RUMMEL

Institut für Astronomische und Physikalische Geodäsie, TU München
(E-mail: rummel@bv.tum.de)

(Received 28 September 2004; Accepted 14 March 2005)

Abstract. Precise global geoid and gravity anomaly information serves essentially three different kinds of applications in Earth sciences: gravity and geoid anomalies reflect density anomalies in oceanic and continental lithosphere and the mantle; dynamic ocean topography as derived from the combination of satellite altimetry and a global geoid model can be directly transformed into a global map of ocean surface circulation; any redistribution or exchange of mass in Earth system results in temporal gravity and geoid changes. After completion of the dedicated gravity satellite missions GRACE and GOCE a high standard of global gravity determination, both of the static and of the time varying field will be attained. Thus, it is the right time to investigate the future needs for improvements in the various fields of Earth sciences and to define the right strategy for future gravity field satellite missions.

Keywords: Geoid, GOCE, GRACE, gravity field, satellite geodesy

After completion of the two dedicated gravity field missions GRACE and GOCE our knowledge of the global gravity field and of its temporal variations will be improved tremendously. The remaining uncertainties (commission errors) in geoid and gravity up to certain spatial scales will be as summarized in Table I. Temporal variations will have been monitored over a time span of five years with a spatial resolution of 200 km and at a mm-level in terms of geoid and 0.05 mGal-level in terms of gravity.

This implies that for future missions the need for improvements beyond these achievements is to be investigated. In (Beutler et al., 2003) the state of art of gravity modelling after GRACE and GOCE has been discussed. It is based on the discussions at a workshop held at the International Space Science Institute (ISSI) at Bern/Switzerland in March 2002. The present collection of articles is focusing on the science issues and on the challenges of future gravity missions. The articles are the outcome of the ESA study *"Enabling Observation Techniques for Future Solid Earth Missions"*. The present introductory article is a modified version of (Rummel, 2003) contained in (Beutler et al., 2003). Improvements could be in terms of geoid and gravity precision over the entire range of spatial and temporal scales, spatial resolution (below 65 km for the static field and below 200 km for the time

TABLE I
Static field: Geoid and Gravity Uncertainties after GRACE and GOCE

Geoid(mm)	Gravity(mGal)	Spatial Scale(km)
1	0.03	200
10	0.2	100
45	2.0	65

varying field), temporal resolution, time span of our record and isolation or identification of individual effects.

Assuming that such improvements are possible, in principle, the following central questions arise:

- What areas of geosciences would profit from such improvements?
- What are the corresponding science issues?
- Is there a certain priority with and sequence in which these issues should be addressed?

Actually, the answer to these questions is the objective of the articles of this issue. A good point of departure are the assessments done for GRACE and GOCE.

In (ESA, 1999) an assessment of the scientific need of improved gravity field knowledge is given from the perspective of the GOCE mission. Thus, in particular the improvement of the static field for geophysics, oceanography, geodesy, glaciology and sea level research has been considered there. The results are summarized in Table II, taken from (ESA, 1999; Table 4.1, p. 80). The science rationale of the GRACE mission is based on a similar exercise, compare (Committee Earth Gravity from Space, 1997). There the emphasis is on the time variable gravity field. The science issues are summarized in Table III.

Three types of applications of the gravity field in earth sciences have to be distinguished. First, any redistribution or exchange of mass in the earth's system results in a change of the earth's gravity field. Thus, the measurement of these temporal variations will serve the analysis of geophysical transport processes. Second, the geoid (the equipotential surface at mean sea level) corresponds to the surface of a hypothetical ocean at rest. The quasi-stationary deviation of the actual ocean surface as measured by altimeters from the geoid, the dynamic ocean topography, can be directly related to ocean surface circulation. Third, the differences between the actual geoid and the surface of a model earth are called geoid anomalies. Analogously the difference between actual gravity and that of such a model earth is denoted gravity anomalies. Both, geoid and gravity anomalies are a measure of the deviation of the actual mass distribution from that underlying the adopted model. The employed Earth model can have any degree of sophistication. The geoid and gravity

TABLE II
Static gravity field, scientific requirements in preparation for GOCE

Application			Accuracy		Spatial resolution
			Geoid (cm)	Gravity (mGal)	Half wavelength-D (km)
Solid earth	Lithosphere/upper mantle density			1–2	100
	Continental lithosphere	Sedimentary		1–2	50–100
		Basins rifts		1–2	20–100
		Tectonic motions		1–2	100–500
	Sesismic hazards			1	100
	Ocean lithosphere/asthenosphere			0.5	100–200
Oceanography	Short scale		1–2		100
			0.2		200
	Basin scale		~0.1		1000
Ice sheets	Rock basement			1–5	50–100
	Ice vertical movements		2		100–1000
Geodesy	Levelling by GPS		1		100–1000
	Unified height systems		1		100–20000
	INS			~1–5	100–1000
	Orbits			~1–3	100–1000
Sea level change			Many of the above applications, with their specific requirements, are relevant to studies of sea level change		

anomalies reflect unmodelled dynamic processes in the oceanic and continental lithosphere and in the upper mantle.

1. Geoid and Gravity Anomalies

Geoid anomalies reflect primarily mass anomalies of large to medium scale spatial extent; gravity anomalies are more related to medium and short scale features. Both, geoid and gravity anomalies, represent density contrasts in the lithosphere and upper mantle. The expectation is that after GOCE geoid anomalies are known with a precision of 1 cm and gravity anomalies with one of 0.1 mGal, both with a spatial resolution of better than 100 km. With future satellite gravity field missions the spatial resolution can probably be increased to 50–60 km. In addition, the global consistency, reliability and precision of the GOCE results could be improved further. This implies that

TABLE III
Time variable gravity field, scientific requirements in preparation for GRACE

Geodynamic effect	Magnitude		Spatial resolution (km)	Main periods
	Geoid (mm)	Gravity (μgal)		
Tides (oceans, soild earth)	100–150		50–5000	Daily, semi-daily, semi-monthly
Atmosphere (IB, NIB, vertical intergation)	15		200–2000	Annual, seasonal, daily, others
Oceans (sea level, currents)	10–15		100–2000	Seasonal, secular
Hydrology (snow, rains, runoff, precipitation, evaporation, reservoirs, ground water)		10	10–1000	Daily to annual
Postglacial rebound		10	1000–10000	Secular
Polar ice and glaciers		5	100–1000	Secular
Soild Earth				
• Earthquakes	0,5		10–100	Single events
• Volcanism	0,5		10–100	Single events
• Tectonics	?		> 500	Secular
• Core and Mantle	?		> 5000	Secular

future missions could, on the one hand, recover missing details, such as the gravity signals due to rift zones or smaller sedimentary basins, and on the other hand, increase the certainty and level of detail of already known gravity features. Very likely, any improvement in resolution beyond 50 km cannot be measured from space. It would have to be deduced from airborne and terrestrial measurements.

At least as important as improvement of global gravity knowledge beyond that of GOCE is its adequate use in geophysical modelling. Earth sciences suffer from a profound lack of direct information about the dynamics and composition of the earth's interior. Everything is derived in an indirect manner. The three primary sources are, most importantly, the analysis of the propagation of seismic waves and, the earth's magnetic and gravity field. All further information is either even more indirect or not globally representative. There is, for example, surface topography, glacial isostatic adjustment, tectonic motion, earth rotation, the laboratory analysis of crust and mantle material and planetology. The trend is towards more and more comprehensive earth modelling, taking most or all of this information into account; global models or models tailored toward specific geophysical phenomena. Seismic tomography has brought a wealth of information about the dynamics of the earth interior. However, due to the nonlinear character of the seismic inversion problem and, in addition, due to the uneven distribution of seismic sources and seismic stations, some fundamental limitations remain, compare e.g. (Snieder and Trampert, 2000). Like seismic inversion, the estimation of the interior density structure from geoid or gravity anomalies is an inverse problem. Its character and its null space are very different, however, from that of seismic tomography. Therefore, the combination of the two could lead to significantly improved earth models. May be the greatest deficiency is currently the uncertainty in the rheological parameters. This aspect is discussed in (King, 2002). The interpretation of geoid anomalies has significantly improved in recent years. Examples are (Cazenave et al., 1986; Simons and Hager, 1997; Lithgow-Bertelloni and Richards, 1998; Kaban et al., 1999; Vermeersen, 2003). Also the joint inversion has made progress, see e.g. (Zerbini et al., 1992) or (Negredo et al., 1999). Nevertheless, global integrated earth modelling in the above sense is still only in an early stage.

2. Dynamic Ocean Topography

Dynamic ocean topography is the quasi-stationary deviation of the actual ocean surface from the geoid (the hypothetical ocean surface at rest). It can be determined from the difference of the actual ocean surface, as measured by satellite altimetry, from an accurate geoid model. It requires the time variable part to be eliminated (by means of averaging and models). Ocean topography is only one to two meters in amplitude. It requires, both, the altimetric surface and the geoid to be determined with centimeter precision. When analyzing future needs, after GRACE and GOCE, several questions have to be addressed:

- Once the geoid could be derived with sub-centimeter precision, can the quasi-stationary ocean surface be determined with the same precision, too?
- What means quasi-stationary? At what temporal and spatial scales does ocean topography change? What are the shortest spatial scales at which a quasi-stationary dynamic topography can still exist? What is their relationship to the Rossby-radius of deformation (It characterizes a spatial-scale boundary. At scales larger than this, the ocean circulation is largely described in terms of geostrophic balance in response to applied forcing (ESA, 1999))?
- How to deal with the so-called omission part of the geoid, i.e. that (short scale) part of the geoid that cannot be sensed from space?

Almost certainly GOCE will not meet all requirements of oceanography in terms of spatial resolution and precision. An ultimate goal could be the determination of the geoid with a precision of 1 cm and with a spatial resolution of shorter than 50 km.

A point of interest could be the translation of the concept of direct measurement of dynamic topography to atmospheric modelling. In oceanography the dynamic topography determines the barotropic flow, see (Open university course team, 1989). The same concept holds for the atmosphere. Atmospheric sounding by GPS yields the geometry of the signal paths along with atmospheric pressure. Together with a global gravity field model the deviations between equipotential and equipressure surfaces could be derived, in principle.

3. Temporal Variations of the Gravitational Field

Gravity variations with time are the result of mass re-distribution in atmosphere, oceans, hydrosphere, glacial areas and within the earth or of mass exchange between these components. Most of these processes are associated to global change phenomena and are essential for the establishment of mass balance within the earth system.

Furthermore, our understanding of the response of our planet to tidal forcing and the study of mass changes in the deep interior (inside the fluid outer core and at the core/mantle boundary) will greatly benefit from the measurement of time changes of the gravity field. Generally, temporal variations are very small; they occur in a wide range of time scales from sudden events, sub-daily, daily, seasonal, to long periodic and secular; they cover all spatial scales, too, from changes in flattening (see e.g. the recent work by Cox and Chao, 2002 and Cazenave and Nerem, 2002), via phenomena related to ocean or continental areas, to regional or local events such as those related to ground water variations of estuaria or to earthquakes, respectively.

The field of measurement of gravity time variations by satellites is largely open. Apart from the analysis of the time changes of the very low zonal harmonics from LAGEOS tracking data, GRACE is the first serious attempt in this direction. The basic challenges are, (1) the small size of the effects to be measured, (2) the broad range of time scales to be covered, (3) the separation of the individual contributions, and (4) aliasing. A medium term strategy must therefore take into consideration the following elements:
- monitoring: a concept based on a series of missions,
- appropriate sampling in space and time: this may lead to the requirement of formation flight of several satellites,
- complementary space missions and data sets needed for the separation of individual contributions and for comprehensive modelling.

4. Planetary Missions

The same concepts of dedicated gravity satellite missions, such as satellite-to-satellite tracking and satellite gradiometry, are applicable to the study of moon and planets, too. Thus, their role for planetary missions should be considered as well. Radar or optical images of surface features (craters, coronae, plateaus, rifts, ridges, roughness versus smoothness), and detailed models of gravity and topography are primary tools of modern planetology. The textbooks by Schubert et al. (2001) or Watts (2001) give an impressive insight into the wealth of information that can be deduced from this information about the current state of our planets.

Yet, the gravity models employed for these investigations exhibit serious deficiencies: Lunar gravity models suffer seriously from the lack of directly observed far side data. Gravity models of Moon, Mars, and Venus suffer to some extent from the spatial variations of the tracking geometry of the connection Earth observatory to planetary orbiter and from the limited variety of orbit parameters of the orbiters. Despite the seemingly high spherical harmonic resolution of the gravity models, their actual significance does often not exceed degree and order 20 or 30, in particular for the moon where far side data is missing. Gravity information from all other planets, planetary sub-satellites and asteroids is derived from fly-byes or orbit perturbations. So far these data only provide the most elementary gravity related information.

5. Conclusions

Once the scientific issues and priorities are identified the questions of their realization by means of space experiments need to be addressed. Already GRACE and GOCE are from the technological point of view extremely challenging missions. Thus, it will not be easy to get to further improvements. In order to build a strategy essentially three variables are at our disposal. These are:
- The choice of the experiment altitude (from extremely high as for lunar ranging or LAGEOS-type satellites to extremely low, even lower than the GOCE orbit) and of the other orbit parameters (in particular inclination and eccentricity). Thereby the experiment altitude is used as a natural low pass filter and the other orbit parameters are employed in particular for the creation of certain sampling characteristics of single satellite or multi-satellite missions.
- The optimization of spectral signal resolution, which is in essence the compensation of field attenuation by means of differential measurements.

This is the core of the gradiometric concept, where the choices of the length of the sensor arm, i.e. the distance between test masses (whether single atoms inside a sensor device or individual satellites) is of fundamental importance.
- The overall performance (precision) of the complete gravity sensor system. This includes considerations concerning the precision of the core sensors (accelerometers, gradiometer, μ-wave link), the required control systems, the (number of) spatial sensor directions, the dynamic range of the system, sampling and de-aliasing strategies.

It will then be the art of finding the optimal synthesis of scientific priorities and the design of mission concepts.

6. Review of Scientific Requirements

In the articles of this issue the need for "post-GRACE + GOCE" gravity improvements will be studied for all fields of Earth sciences and with no prior preference. These fields are geodesy, solid Earth geophysics (from core to crust), hydrology, oceanography, ice sheets and glaciers, sea level, atmosphere, and planetology. The objective is to identify the science issues to be addressed, the corresponding requirements in terms of precision and spatial and temporal resolution, necessary complementary information and their priority.

References

Beutler, G., Rummel, R., Drinkwater, M. R. and von Steiger R.: (eds.), 2003, Earth Gravity Field from Space – from Sensors to Earth Sciences. Space Science Series of ISSI 18, Kluwer Academic Publishers.

Cazenave, A., Dominh, K., Allegre, C. J. and Marsh, J. G.: 1986, *J. Geophys Res. series.* **91B**, 11439–11450.

Cazenave, A. and Nerem, R. S.: 2002, *Sci. Series.* **297**, 783–784.

Committee on Earth Gravity from Space (1997). *Satellite Gravity and the Geosphere.* Washington, D.C: National Academy Press.

Cox, C. M. and Chao, B. F.: 2002, *Sci. Ser.* **297**, 831–833.

ESA: 1999, Gravity Field and Steady-State Ocean Circulation Mission. ESA SP-1233 (1), Noordwijk.

Kaban, M. K., Schwintzer, P. and Tikhotsky, S. A.: 1999, *Geophys. J. Int. Ser.* **136**, 519–536.

King, S. D.: 2002, J Geophys. Res. Ser. 107 B1, ETG 2–1 to 2–10.

Lithgow-Bertelloni, C. and Richards, M. A.: 1998, *Rev. Geophys.* **1**, 27–78 Series 36.

Negredo, A. M., Carminati, E., Barba, S. and Sabadini, R.: 1999, *Geophys. Res. Lett.* **13**, 1945–1948, Series 26.

Open University Course Team (1989). *Ocean Circulation.* Oxford: Butterworth-Heinemann.

Rummel, R.: 2003, in G. Beutler, M. R. Drinkwater, R. Rummel, R. von Steiger, (eds.), *Earth Gravity Field from Space – from Sensors to Earth Sciences.* Space Science Series of ISSI 18, 1–14, Kluwer Academic Publishers.

Schubert, G., Turcotte, D. L. and Olson, P.: 2001, *Mantle Convection in the Earth and Planets*. Cambridge University Press.

Simons, M. and Hager, B. H.: 1997, Nature, series 390, 500–504.

Snieder, R. and Trampert, J.: 2000, *Lect. Notes Earth Sci.* **95**, 93–164 Springer, Berlin.

Vermeersen, B.: 2003, in G. Beutler, M. R. Drinkwater, R. Rummel and R. von Steiger (eds.), *Earth Gravity Field from Space – from Sensors to Earth Sciences*. Space Science Series of ISSI 18, 105–113, Kluwer.

Watts, A. B.: 2001, *Isostasy and Flexure of the Lithosphere*. Cambridge University Press.

Zerbini, S., Achache, J., Anderson, A. J., Arnet, F., Geiger, A., Klingelé, E., Sabadini, R. and Tinti, S.: 1992, Study of Geophysical Impact of High-Resolution Earth Potential Field Information. ESA-study, final report.

Earth, Moon, and Planets (2005) 94: 13–29
DOI 10.1007/s11038-005-3756-7

FUTURE SATELLITE GRAVIMETRY FOR GEODESY

J. FLURY and R. RUMMEL

Institut für Astronomische und Physikalische Geodäsie, TU München
(E-mail: flury@bv.tum.de; rummel@bv.tum.de)

(Received 4 October 2004; Accepted 14 March 2005)

Abstract. After GRACE and GOCE there will still be need and room for improvement of the knowledge (1) of the static gravity field at spatial scales between 40 km and 100 km, and (2) of the time varying gravity field at scales smaller than 500 km. This is shown based on the analysis of spectral signal power of various gravity field components and on the comparison with current knowledge and expected performance of GRACE and GOCE. Both, accuracy and resolution can be improved by future dedicated gravity satellite missions. For applications in geodesy, the spectral omission error due to the limited spatial resolution of a gravity satellite mission is a limiting factor. The recommended strategy is to extend as far as possible the spatial resolution of future missions, and to improve at the same time the modelling of the very small scale components using terrestrial gravity information and topographic models. We discuss the geodetic needs in improved gravity models in the areas of precise height systems, GNSS levelling, inertial navigation and precise orbit determination. Today global height systems with a 1 cm accuracy are required for sea level and ocean circulation studies. This can be achieved by a future satellite mission with higher spatial resolution in combination with improved local and regional gravity field modelling. A similar strategy could improve the very economic method of determination of physical heights by GNSS levelling from the decimeter to the centimeter level. In inertial vehicle navigation, in particular in sub-marine, aircraft and missile guidance, any improvement of global gravity field models would help to improve reliability and the radius of operation.

Keywords: Geodesy, gravity field, heights, GNSS levelling, inertial navigation

1. Introduction

All currently available geoid or gravity models are based on satellite orbit analysis, satellite radar altimetry in ocean areas, terrestrial and shipborne gravimetry and topographic models. Long-term observation and analysis resulted in global gravity field models such as the Earth Gravity Model EGM96 (Lemoine et al., 1998). Despite of the wealth of input data and sophistication of the computational processes these models are still rather heterogeneous in terms of resolution and accuracy. At present, a major step forward in gravity field knowledge is achieved by a first generation of dedicated gravity field satellite missions, CHAMP (Reigber et al., 2003), GRACE (Tapley et al., 2004b) and GOCE (ESA, 1999).

In geodesy the requirements of geoid and gravity field are particularly high. Highly accurate and homogeneous gravity field information is needed for the establishment of unified global height systems and for the calculation of physical heights from ellipsoidal GNSS heights. Precise and homogeneous height systems are important for engineering purposes as well as for Earth sciences, e.g. for sea level studies. With GRACE and GOCE, considerable improvements will be achieved, but the requirements in terms of accuracy and spatial resolution will not be fully met. Inertial navigation and precise orbit determination are other geodetic fields requiring very accurate gravity field knowledge.

A new and very challenging field is the observation and analysis of the time variable gravity field due to mass variations in atmosphere, oceans, continental water cycle, ice covered areas and solid Earth (Committee, 1997; Ilk et al., 2005). The satellite mission GRACE enables a first view on this field, but many open questions will remain. With GRACE, and even more with future satellite gravity missions, temporal gravity field variations will become an important subject of geodetic research. For adequate analysis and modelling, a close cooperation between geodesy, geophysics, oceanography, glaciology, hydrology and other Earth sciences is mandatory.

There exist reliable mathematical "rules" about the average behaviour of the gravity field and geoid under the assumption of stationarity and isotropy. These rules are power spectra of average signal power, expressed in terms of spherical harmonic degree variances or degree rms. We use such power spectra as a starting point for a discussion of the state-of-the-art, of shortcomings and future needs in gravity field knowledge, cf. Section 2.1. At short wavelengths gravity information is insufficient or non-existent. For wavelengths smaller than 100 km, this situation will remain unchanged even after the completion of the first generation of dedicated gravity field satellite missions. Statistically, the unknown small-scale gravity signal is dealt with as omission error, cf. Section 2.2. Much more uncertain than the stationary gravity field characteristics is the signal size and behaviour of the time variable gravity field, cf. Section 2.3. The stationary and time variable gravity field signal amplitudes and length scales give the background for the subsequent analysis of future geodetic requirements and possible improvements by future satellite missions, cf. Section 3.

2. Gravity Field and Geoid: State-of-the-Art

2.1. STATIONARY GEOID

In the following we discuss the signal amplitudes of the Earth's gravity field at various wavelengths and spatial scales, respectively. The discussion is based on Figure 1 where degree rms values from spherical harmonic geoid expansions are shown. The figure also explains where the current geoid information comes from and where there is room and need for improvement.

- The two central lines are those denoted "Tscherning-Rapp" and "Kaula". They represent models of the geoid signal size (in meters) as a function of spherical harmonic degree or average spatial scale (half-wavelength in kilometers). They are based on satellite and terrestrial data translated into a simple mathematical expression (rule of thumb). At small length scales these curves are extrapolated and pure speculation. In addition, they assume homogeneous signal characteristics all over the globe. Nevertheless, the two rules give an excellent impression of the overall (logarithmic) signal strength decrease with increasing spherical harmonic degree.
- The line for the Earth Gravity Model EGM96 shows the signal decrease of the geoid based on real data. It is derived from satellite, terrestrial and altimetric gravity information. Its maximum spherical harmonic degree is 360 (about 60 km). It fits well between the Tscherning-Rapp and Kaula lines.

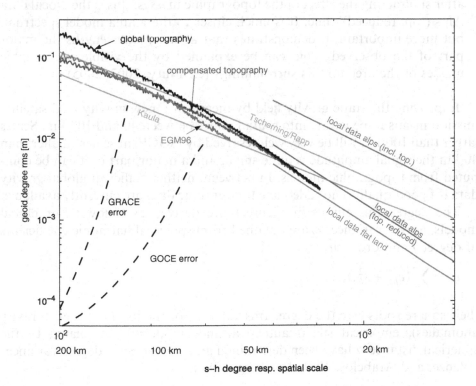

Figure 1. Single content of the static gravity field, shown between spherical harmonics degree 100 and 2000, in terms of geoid signal degree rms-values, as well as GRACE and GOCE error rms-values. See the explanation in the text.

– The "global topography" line is based on a geoid computation from the gravity potential of all visible topographic masses (and ocean depths) assuming constant density. It seriously overestimates the geoid signal strength because it neglects any isostatic compensation of topographic masses.
– Therefore, in a second computation, mass compensation has been taken into account using a simple Airy compensation model. Now the "compensated topography" line fits well to the measured EGM96 line. One can also observe that at smaller length scales, say below 50 km, topographic masses are not compensated anymore according to this model.
– From the GRACE and GOCE error curves one can deduce the state-of-art of geoid knowledge after GRACE and GOCE, respectively.
– At short wavelengths (10–40 km) two representative areas were selected with very good terrestrial gravity data coverage: a local test area in the Alps, which results in the geoid spectral line "local data Alps" and an area of similar size but flat, resulting in the spectral line "local data flat land". One observes the large difference in amplitude (one order of magnitude).
– Finally, for the Alpine test area, the geoid spectrum has been computed after subtracting the effect of the topographic masses; this is the "local data Alps (top. reduced)" line. It is much closer to the Kaula model spectrum, but more important, it demonstrates that at short wavelengths the major part of the observed geoid can be explained by the visible topographic masses of the area and its surroundings (cf. Flury, 2002, 2005).

Improving the static gravity field by means of a new gravity field satellite mission means to penetrate into spatial scales between 40 and 100 km. Scales larger than 100 km will be very well resolved by GOCE; at scales smaller than 40 km the signal amplitude is quite small, and a major part of it can be computed from topographic models. In between, neither sufficient global gravity data (of good quality) nor adequate topographic data are currently available.

The degree rms curves in Figure 1 are derived as follows: For global models, degree variances c_l are obtained from spherical harmonic coefficients of degree l and order m

$$c_l = \sum_m (c_{lm}^2 + s_{lm}^2),$$

their square roots give the degree rms values. For the local data sets (gravity anomalies), empirical signal autocovariance models $C(\psi)$ related to the spherical distance ψ have been determined and converted to degree variances (Wenzel and Arabelos, 1981; Forsberg, 1984):

$$c_l = \int_{\psi=0}^{\Psi} C(\psi) P_l(\cos \psi) \sin \psi d\psi.$$

$P_l(\cos \psi)$ are Legendre polynomials. The integration is performed up to an appropriate maximum distance ψ. The gravity anomaly degree variances obtained from local data can be represented by simple power laws (cf. Flury, 2005) and converted to geoid heights. The GRACE and GOCE error models are obtained from mission simulations, cf. Sneeuw et al., this issue.

2.2. GEOID OMISSION PART

Any satellite gravity field mission will measure the global geoid with a certain precision and a certain spatial resolution. The size of the signal not resolved by the measurements constitutes the so-called omission part. For most geophysical science applications a certain finite spatial resolution is sufficient, corresponding e.g. to the grid spacing of a finite element model. Then the omission part does not enter into considerations as long as the geoid resolution fits to the model resolution. For some other applications, the full geoid information is needed at individual points on earth, i.e. the full spectral content from zero to infinity. In those cases the omission part enters into the error budget. The omission part can be reduced by improving the spatial resolution of a space mission or by adding local geoid information as derived e.g. from local gravity surveys and topographic data.

Figure 2 shows the geoid omission part in centimetres as a function of spherical harmonic degree in the window between degree 300, which represents the situation after GOCE, and degree 700, which a future mission could possibly resolve, corresponding to the range of 70–30 km half-wavelength, respectively. The full omission error, shown by the black line, is still rather high, decreasing from 28 cm at degree 300 to 10 cm at degree 700. It should be noted, however, that by adding local gravity information in a circle of radius 0.5°, 1°, 2° or 5° (50 km, 100 km, 200 km, 500 km) around the computation point the omission error can be reduced significantly. The error for a 0.5° radius cap is of special interest: such an area could – even if no local data were available – be filled up with about 100 terrestrial gravity measurements very fast and economically in sufficient accuracy.

2.3. TIME VARIABLE GRAVITY FIELD

The geoid and gravity are changing with time due to the tides of sun, moon and planets, due to mass movements and mass exchange in Earth system and, as a secondary effect, due to deformation of the Earth's surface as a consequence to such mass motions. The changes are small, ranging from below a

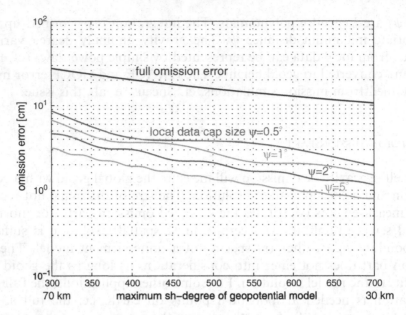

Figure 2. Omission error modeled using Tscherning/Rapp degree variance model. For each spherical harmonic degree the omission part up to infinity is shown, assuming that the signal up to this degree is covered by a geopotential model. The omission error reduces greatly when local gravity data for a relatively small cap size is added, with spherical cap radii of 0.5°, 1°, 2° or 5°.

mm to a few dm for the geoid, they occur at all time scales from secular to sudden and at all spatial scales from global to very local. From terrestrial gravity measurements the global pattern of Earth tides is known rather well; also temporal changes due to postglacial rebound or secular crustal movements are nowadays well measurable. All other effects, such as changes in groundwater level or ocean or atmospheric loading are so far difficult to identify and quantify by *in-situ* measurements. From satellite orbit analysis time variations in the very low degree zonal spherical harmonic coefficients can be determined rather well. Their physical interpretation proves difficult, however (Cazenave and Nerem, 2002; Cox and Chao, 2002). The situation is expected to improve significantly with the monthly sets of spherical harmonic coefficients from GRACE, from which changes in atmospheric pressure, ocean bottom pressure or in the hydrological cycle will be derived, compare Wahr et al. (1998) and Tapley et al. (2004a, b).

Like it is done for the static gravity field, it is common practice to represent the signal strength of the various contributions of temporal geoid variations in terms of signal degree variances. On the one hand, this allows to compare the spectral signal characteristics of the individual geophysical signals; on the other hand, one can compare the various signals with the expected spectral characteristics of the measurement noise of space missions and deduce thereof indications about their observability. There are two difficulties with this approach. First, for some of the geophysical signals knowledge about their temporal and spatial behaviour is rather poor; consequently the degree variance lines may be unrepresentative. Second, some of the considered phenomena are confined to land or ocean or certain geographical regions, which makes the use of spherical harmonic degree variances somewhat problematic. Thus, the spectra are to be seen with a certain "grain of salt".

In essence, the temporal variations can be divided into two classes: those that need to be studied by future gravity field satellite missions and those that are to be considered as "disturbances". A special class of disturbances arises from periodic, high frequency geophysical phenomena that map as "alias" into the spectral range of interest due to the peculiar space-time sampling of a satellite. In particular, semi-diurnal and diurnal phenomena such as the tides of the solid Earth, oceans and atmosphere belong to this category.

Figures 3 and 4 give an impression of the spectral geoid signal of some geophysical effects. Comparison with the geoid degree rms-values of the

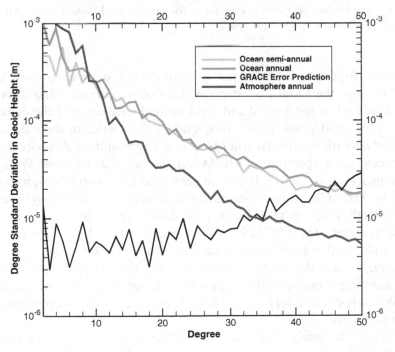

Figure 3. Geoid degree rms-values of the annual variations in atmospheric density of the annual and semi-annual ocean mass changes. For comparison, the expected GRACE noise spectrum is added.

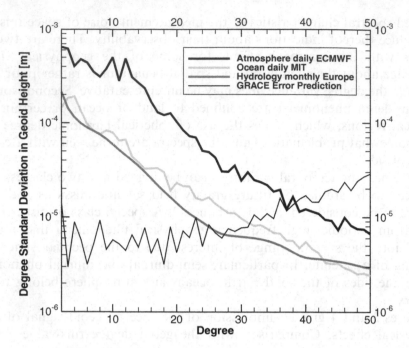

Figure 4. Geoid degree rms-values of the daily variations in atmospheric density, the daily mass changes of the oceans monthly changes in the hydrological cycle. For comparison, the expected GRACE noise spectrum is added.

static field, Figure 1, explains the smallness of the time variable signals. Figure 3 shows the geoid degree rms-values of the annual variations of the atmosphere and of the annual and semi-annual variations of the oceans. All three spectra are derived from corresponding time series of daily data. Their uncertainty is rather large. In order to get an impression of their observability the expected noise spectrum of GRACE is included as well (see Wahr et al., 1998; Wünsch et al., 2001). It can be seen that time variable signals can be derived by GRACE up to about degree and order 35, or at length scales of about 500–600 km. Figure 4 shows some daily and monthly geoid variations (daily atmosphere from ECMWF, daily ocean from MIT ocean circulation model, and – with a high uncertainty – monthly changes in the hydrological water cycle). These three time series may cause aliasing problems, depending on the sampling strategy of future gravity field satellite missions (see Gruber, 2001; Wünsch et al., 2001). The GRACE expected noise spectrum is added for completeness.

Based on this analysis of the signal behaviour of stationary and time-variable gravity and geoid and the current state-of-the-art of commission and omission error one can now turn to gravity and geoid applications in geodesy and (in the following articles) in Earth sciences.

3. Geodetic Requirements

It is common to all applications of gravity and geoid information in geodesy, mapping, geomatics and engineering that point values or differences of point values are needed. The only exception is orbit determination. This implies that the band limited geoid and gravity information deduced from satellite measurements is only part of the required total geoid and gravity point values. The missing complementary, small scale part has to come from local airborne, shipborne or terrestrial gravimetric surveys and from topographic models. After GOCE this is the signal part above spherical harmonic degree 250, see Rummel (2004, this issue). If the small scale geoid and gravity contribution is neglected, a rather high omission error has to be accepted, as was discussed in Section 2. While after GOCE the geoid error at degree and order 250 will amount to a few cm only, the omission part is about 30 cm 1 σ error.

Thus, what would be the proper strategy in these fields for the future?

(1) An improvement of GOCE precision without any improvement in resolution would reduce the geoid commission error from a few cm after GOCE to the sub-cm level. However, this will not result in major breakthroughs in view of the large omission error.

(2) An improvement in spatial resolution from a maximum degree of 250 to, say, degree 700 (or 30 km) would decrease the remaining omission error from 30 to 10 cm. This would help considerably, but it may prove very difficult to get such high resolution from space techniques.

(3) Reduction of the omission error by means of local gravimetric surveys and topographic models: In a concerted effort – based on new, well controlled measurement campaigns, re-analysis of existing data sets and new high resolution terrain models – the local omission error can be reduced to the cm-level.

As a consequence, the strategy must be: (a) Extension of the spatial resolution of future dedicated satellite missions, beyond degree 250 without loss of accuracy, and (b) reduction of the omission error for selected regions and applications to the cm-level.

3.1. HEIGHTS

In most countries, well defined and official "heights above sea level" are made available and maintained for use in engineering, mapping, cadastre, geo-referencing as well as for use in science. These are so-called physical heights which means they carry information about the direction of flow of water. National height systems usually refer to sea level and are tied to a tide gauge at an arbitrarily selected point. This results in offsets between height systems at borders and across oceans or ocean straits. For Europe, the offsets are in the order of some decimetres and are rather well known, with an accuracy of 1–10 cm (Figure 5, from Sacher et al., 1999). For some continents they are unknown and may be much larger. Physical heights are derived by spirit levelling plus gravimetry. This approach has the tendency to accumulate systematic errors that may amount to several cm or even dm on a continental scale (Haines et al., 2003), in addition to the tide gauge offsets.

Ideally, "heights above sea level" should be "heights above the geoid", i.e. in an ideal case all height systems should refer to one and the same level surface. Only then physical heights all over the earth would become comparable, which would lead to the establishment of a world height reference system. One would be able to decide whether an arbitrary coastal

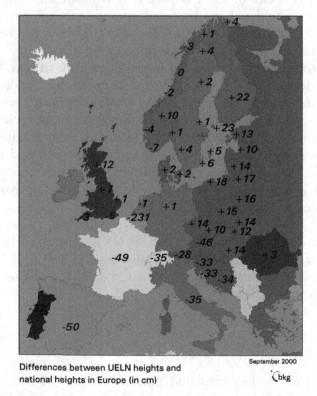

Differences between UELN heights and national heights in Europe (in cm)

September 2000

Figure 5. Offsets between national height systems in Europe (in cm), from Sacher et al. (1999). Countries in the same colour are connected to the same datum point (tide gauge). By courtesy of Bundesamt für Kartographie und Geodäsie, Frankfurt.

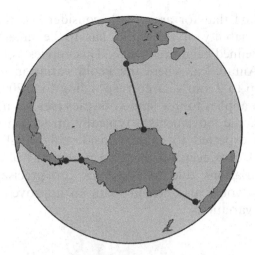

Figure 6. Examples for links between tide gauges (crossing the Antarctic Circumpolar Current), for which high precision geoid differences are required for ocean circulation modelling.

point in Australia would be higher or lower than a second point somewhere in Europe. This is of some importance for regional and global mapping, geo-referencing and geo-information, and it is of great importance for global sea level and ocean circulation studies. Tidal records would become comparable globally, too (see Figure 6). The required geoid accuracy is below 10 cm (commission plus omission part) or on the long term below 1 cm. Official institutions like the US National Geospatial-Intelligence Agency NGA strive for a continental or global accuracy of 2 cm. After GOCE, a height system accuracy of approximately 5–10 cm – commission and omission error – will be achievable for all areas where a sufficient coverage of local gravity data is available to deal with the omission part (see Section 2; Arabelos and Tscherning, 2001; Rummel, 2002). For a 1 cm accuracy, however, one has to take into account local data up to much larger distances, and a satellite mission with higher resolution is needed.

For engineering purposes (construction of bridges, long tunnels, canals) precise physical height differences and deflections of the vertical are needed. The local character of these applications emphasizes the need for very precise local gravimetric and topographic information and relaxes the requirement for very precise global (satellite derived) geoid and gravity models. However, for these applications an increased resolution of the geopotential models would lead to cost reduction in the sense that less effort would be required to check and refine the geoid and height reference locally, e.g. that the amount of required local gravity observations would decrease substantially.

It has to be noted that for areas with considerable vertical land movements the time variability of the geoid has to be taken into account to maintain the well defined height reference. This applies especially to Canada, Scandinavia and Antarctica, where the geoid variation due to postglacial rebound reaches up to 2 mm/year corresponding to a 20 cm change over a century, but it also applies to low land countries (see Vermeersen, 2004, this issue). Vertical tectonic movements – typically up to some mm/year (Lambeck, 1988)   are expected to cause considerably smaller geoid changes (Marti et al., 2003). However, on the long run it may also be necessary to consider geoid variations due to tectonics for precise height systems. Therefore, geodesy would also benefit from an improved determination of secular geoid time variations.

3.2. GNSS LEVELLING

Nowadays, by means of GPS and other space positioning methods, absolute point positions or position differences can be derived at the cm-level or sub-cm level, respectively, depending on the sophistication of the measurement setup. There is a clear trend to even higher precisions once the next generation of satellite navigation systems with more precise clocks as well as better atmospheric monitoring become available. 3-D point positions or position differences can be directly transformed into geographical coordinates or coordinate differences and ellipsoidal heights or height differences. In combination with a precise geoid model, ellipsoidal heights can be conveniently translated into physical heights (heights above the geoid). This technique circumvents the tedious, time-consuming traditional geodetic levelling and avoids the systematic errors inherent to this method. Precondition is, however, the availability of absolute geoid heights or height differences with a precision comparable to that of the ellipsoidal heights derived by means of GNSS. This is to say the required geoid precision (commission and omission part) has to be at the cm or sub-cm level.

Already now the technique of GPS-levelling is applied worldwide at a much more moderate level of accuracy. The fields of application are engineering (construction of streets, bridges, tunnels, canals), mapping and geo-information, exploration and many more. The method is the only economic way to get a well defined height reference in unsurveyed areas, be it in developing countries or in the polar regions. It is of special interest for areas where permanent height benchmarks cannot be built, e.g. for inundation zones, where the benchmarks get destroyed by every flood.

Where no local gravity data in good quality are available, GNSS levelling is affected by the full geoid error (commission and omission) of a geopotential

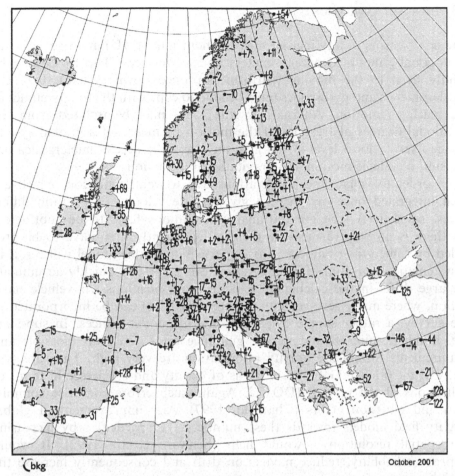

bkg   October 2001

Differences between EUVN GPS/levelling geoid and EGG97 geoid (in cm)

Figure 7. Errors for GNSS Levelling using the geoid model EGG97 (Denker and Torge, 1998), based on EGM96 geopotential model and a very large terrestrial gravity data base. By courtesy of Bundesamt für Kartographie und Geodäsie, Frankfurt.

model. This error will be – depending on the distance between stations – at a few cm after GOCE. It could be lowered down to the centimeter level by a satellite mission resolving spatial scales of the geoid up to spherical harmonic degree 700.

Figure 7 (from Ihde et al., 2002; see also Denker and Torge, 1998) shows today's state-of-the-art, where even in a well surveyed area like Europe the geoid errors amount to several dm in general, in some cases even to more than 1 m.

3.3. INERTIAL NAVIGATION

The core sensors of any inertial measurement unit (IMU) are a set of three orthogonally mounted accelerometers and gyroscopes. The IMU may be rigidly fixed to the moving platform (strapped down system); then the accelerometers and gyroscopes must be able to cope with the full dynamics of the platform motion. Alternatively the IMU may be isolated from the rotational degrees of freedom of the motion by means of a gimbal system (space fixed or local level system). In any case, the accelerometers measure the sum of vehicle motion and gravitational attraction.

In order to isolate the accelerations due to vehicle motion, which are then integrated once or twice to give vehicle velocity or position differences, respectively, the gravitational accelerations have to be subtracted. As there is no way to measure them independently they have to be provided by some gravity model. Any imperfection in this model will result in a systematic error, and after integration this error will quickly accumulate to large drifts in the calculated velocities and positions. In vehicle navigation, where many zero-velocity updates (ZUPT's) can be incorporated in the survey, a rather simple ellipsoidal gravity model suffices. In sub-marine, borehole, aircraft and missile guidance no ZUPT's are possible and requirements for precise gravity information are very high. They are typically at the 0.1–1 mGal level in terms of gravity and 0.1 arcsec in terms of deflections of the vertical (DOV's). Again, such errors comprise commission and omission errors (Chatfield, 1997). Any improvement of global gravity field models towards these numbers – in particular by increasing the spatial resolution – would help (Schwarz et al., 1992). It would improve reliablility, reduce navigation drift and consequently increase the radius of operation.

3.4. SATELLITE ORBIT DETERMINATION

After GRACE and GOCE a very good Earth gravity model will be available for the determination of satellite orbits. However, even then a gravity model based on only one or two missions may exhibit some specific weaknesses. This can be seen when using the current CHAMP only solutions for orbit determination of other satellites (Schrama, 2003). Complementary missions may therefore still prove to be of importance, in particular missions at higher altitude or with different orbit inclinations. For this, inexpensive missions of the type LAGEOS or high orbiting satellites equipped with continuous tracking devices such as GPS receivers could serve. For the determination of low zonal coefficients and their secular variations, long mission durations (>10 years) may remain important.

On the long term, from a perfect knowledge of the gravity field in conjunction with orbits derived from high–low SST using a GNSS (e.g. using the kinematic method), the influence of the non-gravitational forces could be studied, which could be of value for atmosphere physics. In summary, there is no immediate need from the point of view of orbit determination for a follow-on dedicated gravity mission.

4. Conclusions

Geodesy, including navigation, mapping, engineering and geo-referencing requires geoid, gravity anomaly or deflections of the vertical (DOV) values in the absolute sense. Thus, not only the (commission) error of these quantities deduced from potential future satellite gravity missions has to be taken into account. At least as important is the reduction of the omission error, i.e. the signal part that cannot be resolved from space.

A medium (and long) term accuracy goal (commission and omission part) is:

- 10 cm (1 cm) for geoid heights,
- 1 mGal (0.1 mGal) for gravity anomalies,
- 1 arcsec (0.1 arcsec) for DOV's.

The corresponding strategy must be:

- Without loss of precision extend the spatial resolution of a future gravity satellite mission from sperical harmonic degree $l_{max} = 250$ (corresponding to 80 km half-wavelength) to $l_{max} = 400$ (50 km) or may be even $l_{max} = 700$ (35 km).
- Reduction of the remaining omission part by means of local gravimetric surveys (airborne, shipborne, terrestrial), re-analysis of existing gravity data sets and new high resolution digital terrain models.

Acknowledgements

This work was funded in part by Deutsches Zentrum für Luft- und Raumfahrt (DLR) which is gratefully acknowledged.

References

Arabelos, D. and Tscherning, C. C.: 2001, *J. Geodesy*. **75**, 308–312.
Cazenave, A. and Nerem, R. S.: 2002, *Science*. **297**, 783–784.

Chatfield, A. B.: 1997, Fundamentals of high accuracy inertial navigation, in *Progress in Astronautics and Aeronautics*, Vol. 174, AIAA Reston.

Committee on Earth Gravity from Space (1997). *Satellite Gravity and the Geosphere*. Washington, D.C: National Academy Press.

Cox, C. M. and Chao, B. F.: 2002, *Science*. **297**, 831–833.

Denker, H. and Torge, W.: 1998, The European Gravimetric Quasigeoid EGG97, in *International Association of Geodesy Symposia*, Vol. 119, Geodesy on the Move, Springer.

ESA, European Space Agency: 1999, Gravity Field and Steady-State Ocean Circulation Mission (GOCE), Report for mission selection, SP-1233 (1), Noordwijk.

Flury, J.: 2002, Schwerefeldfunktionale im Gebirge: Modellierungsgenauigkeit, Messpunktdichte und Darstellungsfehler am Beispiel des Testnetzes Estergebirge, Deutsche Geodaetische Kommission, Series C 557, München.

Flury, J.: 2005, J. Geodesy, in revision.

Forsberg, R.: 1984, Local covariance functions and density distributions, Dep. of Geodetic Science Report 356, Ohio State Univ. Columbus.

Gruber, Th.: 2001, Identification of Processing and Product Synergies for Gravity Missions in View of the CHAMP and GRACE Science Data System Developments, in *Proceedings of 1st International GOCE User Workshop*, ESA Publication Division, Report WPP-188.

Haines, K., Hipkin R., Beggan C., Bingley R., Hernandez F., Holt J., Baker T. and Bingham R. J.: 2003, in G. Beutler, M. Drinkwater, R. Rummel and R. von Steiger (eds.), *Earth Gravity Field from Space: From Sensors to Earth Sciences*, Space Sciences Series of ISSI Vol. 18, Kluwer, pp. 205–216.

Ihde, J., Adam J., Gurtner W., Harsson B. G., Sacher M., Schlüter W. and Wöppelmann G.: 2002, The Height Solution of the European Vertical Reference Network (EUVN), in *EUREF-Publication Nr. 11/I* pp. 53–70, Mitteilungen des Bundesamtes für Kartographie und Geodäsie, 25, Frankfurt am Main.

Ilk, K. H., Flury, Rummel R., Schwintzer P., Bosch W., Haas C., Schröter J., Stammer D., Zahel W., Miller H., Dietrich R., Huybrechts P., Schmeling H., Wolf D., Götze H.J., Riegger J., Bárdossy A., Güntner A. and Gruber T.: 2005, *Mass Transport and Mass Distribution in the Earth System*, Contribution of the new generation of satellite gravity and altimetry missions to geosciences, 2nd edn, GOCE Projektbüro TU München, GFZ Potsdam.

Lambeck, K.: 1988, *Geophysical Geodesy: The Slow Deformation of the Earth*. Oxford University Press.

Lemoine, F. G., Kenyon S. C., Factor J. K., Trimmer R. G., Pavlis N. K., Chinn D. S., Cox C. M., Klosko S. M., Luthcke S. B., Torrence M. H., Wang Y. M., Williamson R. G., Pavlis E. C., Rapp R. H. and Olson T. R.: 1998, The development of the joint NASA GSFC and the National Imagery and Mapping Agency (NIMA) geopotential model EGM96. NASA Technical Paper NASA/TP-1998-206861, Goddard Space Flight Center, Greenbelt.

Marti, U., Schlatter A. and Brockmann E.: 2003, Analysis of vertical movements in Switzerland, Presentation EGS-AGU-EUG general assembly Nice 2003.

Reigber, C.Schwintzer, P.Neumayer, K.-H.Barthelmes, F.König, R.Förste, C.Balmino, G.Biancale, R.Lemoine, J.-M.Loyer, S.Bruinsma, S.Perosanz, F. and Fayard, T.: 2003, *Adv. Space Res.* **31**(8), 1883–1888.

Rummel, R.: 2002, Global unification of height systems and GOCE, in M. Sideris (eds.), *Gravity, geoid and geodynamics*, IAG symposium Banff 2000, Springer, pp. 13–20.

Rummel, R.: 2004, *Earth, Moon and Planets*, this issue.

Sacher, M., Ihde, J. and Seeger, H.: 1999, Preliminary Transformation Relations between National European Height Systems and the United European Levelling Network (UELN), in *Report on the Symposium of the IAG Subcommission for Europe (EUREF)*, pp. 80–86,

Prague, 2–5 June 1999, Veröffentlichung der Bayer. Komm. für die Internationale Erdmessung, München.

Schwarz, K. P., Colombo, O., Hein, G., Knickmeyer, E. T.: 1992, Requirements for airborne vector gravimetry, in *From Mars to Greenland*, IAG symposium 1991, Springer, pp. 273–283.

Schrama, E. J. O.: 2003, in: G. Beutler, M. Drinkwater, R. Rummel and R. von Steiger (eds.), *Earth Gravity Field from Space: From Sensors to Earth Sciences*, Space Sciences Series of the ISSI 18, Kluwer, pp. 179–194.

Tapley, B. D., Bettadpur, S., Ries, J. C., Thompson, P. F. and Watkins, M. M.: 2004a, *Science* **305** 503–505.

Tapley, B. D., Bettadpur, S., Watkins, M. and Reigber, C.: 2004b, *Geophys. Res. Lett.* 31, L09607.

Vermeersen, B.: 2004, *Earth, Moon Planets*, this issue.

Wahr, J.Molenaar, M. and Bryan, F.: 1998, *J. Geophys. Res.* **103**(B12), 30205–30230.

Wenzel, H. G. and Arabelos, D.: 1981, *Zeitschrift für Vermessungswesen.* **106**, 234–243.

Wünsch, J.Thomas, M. and Gruber, Th.: 2001, *Geophys. J. Int.* **147**, 28–434.

Earth, Moon, and Planets (2005) 94: 31–40
DOI 10.1007/s11038-004-6816-5

CHALLENGES FROM SOLID EARTH DYNAMICS FOR SATELLITE GRAVITY FIELD MISSIONS IN THE POST-GOCE ERA

BERT L. A. VERMEERSEN

DEOS, Delft University of Technology, Kluyverweg 1, 2629 HS, Delft, The Netherlands
(E-mail: b.vermeersen@lr.tudelft.nl)

(Received 11 August 2004; Accepted 25 November 2004)

Abstract. Examples from four main categories of solid-earth deformation processes are discussed for which the GOCE and GRACE satellite gravity missions will not provide a high enough spatial or temporal resolution or a sufficient accuracy. Quasi-static and episodic solid-earth deformation would benefit from a new satellite gravity mission that would provide a higher combined spatial and temporal resolution. Seismic and core periodic motions would benefit from a new satellite mission that would be able to detect gravity variations with a higher temporal resolution combined with very high accuracies.

Keywords: Deformation, gravity, resolution

1. Introduction

In solid-earth research after the GOCE mission the interest in temporal variations of the gravity field will be increasing. As far as effects near the Earth's surface are concerned, many requirements will not be met by the spatial resolution of temporal variations observed by GRACE.

Three main areas of solid-earth geodynamics can be distinguished on which solid-earth dynamics would benefit from a "GRACE-like", or even better, temporal resolution for "GOCE-like" spatial scales: post-glacial rebound and concomitant sea-level variations; co- and post-seismic solid-earth deformation; and mantle convection and plate tectonics.

The situation is different for the detection of core motions and seismic modes, where the spatial resolution is not critical for core motions, but a very high accuracy and temporal resolution are required.

In the following sections, each of these four geodynamical processes and their geodetic signatures are briefly introduced, after which an overview is given of what GRACE and GOCE are expected to contribute to each of these fields, and where GRACE and GOCE fall short.

It should be noted that gravity effects from crustal displacements due to hydrological, oceanic and atmospheric loadings are not treated here.

2. Glacial Rebound and Associated Sea Level Variations

Glacial Isostatic Adjustment (GIA) of the solid Earth due to the waxing and waning of Late-Pleistocene Ice-Age cycles has created geoid and gravity anomalies, although the separation between GIA-induced contributions and those induced by plate tectonics and mantle dynamics is not always obvious. For example, it is now widely acknowledged that the deep geoid low above Canada is partly due to non-GIA induced lithosphere and mantle hetero-geneities and partly attributable to GIA (e.g., Simons and Hager, 1997).

Whereas the geoid above Canada is related to (at least) two geody-namical processes, it is thought that secular geoid and gravity anomaly variations are only triggered by post-glacial rebound (e.g., Wahr and Davis, 2002). Figure 1, taken from Wahr and Davis (2002), shows in the top panel the secular degree geoid amplitude as function of spherical harmonic degree for expected GRACE errors and predictions of three models: GIA; present-day Antarctic ice decay equivalent with a sea-level

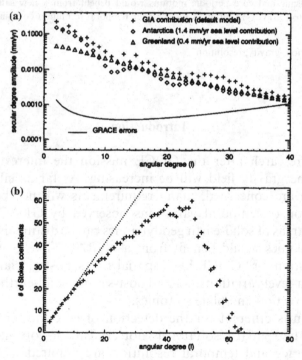

Figure 1. (a) Present-day secular geoid change model predictions for GIA due to Late-Pleis-tocene deglaciation with a default model (see text for details), together with the effects of the maximum contemporary melt scenarios for the Antarctic and Greenland ice caps. The solid line represents the predicted errors in the GRACE data; (b) Total number of Stokes coeffi-cients as function of spherical harmonic degree for which the GIA signal of *a* is expected to be larger than the expected secular GRACE measurement error (*figure taken from Wahr and Davis, 2002*).

rise of 1.4 mm/yr; and present-day Greenland ice cap decay equivalent with a sea-level rise of 0.4 mm/yr.

The default model for the GIA contribution has an earth model with a lithospheric thickness of 120 km and upper and lower mantle viscosities of 10^{21} and 10^{22} Pa s, respectively, while the radial elastic and constitutional parametrization is based on the Preliminary Reference Earth Model (PREM) by Dziewonski and Anderson (1981) and the Late-Pleistocene ice-decay model is based on ICE-3G by Tushingham and Peltier (1991). From this top panel of Figure 1 it can be derived that GRACE is expected to be able to discern GIA and (maximum) present-day Antarctic and Greenland ice-cap variations up to about harmonic degree 40. For degrees between about 40 and 60, still a number of Stokes coefficients related to the defaults GIA model are expected to become detectable by GRACE, as is shown in the lower panel of Figure 1. In total, for all harmonic degree up till degree 60 about 2000 Stokes coefficients should become detectable. It should be noted here that the pre-launch estimates of GRACE as given in Figure 1 are considerably higher than what is achieved in the most recent models. For instance, compare with Figure 1 of Tapley et al. (2004), taking into account that the GRACE uncertainty estimates in Figure 1 are based on a 5-year mission length.

Figure 2, also taken from Wahr and Davis (2002), shows the sensitivity of mantle viscosity and lithospheric thickness variations in GIA models with respect to secular GRACE measurement errors. The data points represent differences between two lower mantle viscosity models "v_{LM}", two upper mantle viscosity models "v_{UM}" (viscosities given in Pa·s) and two lithospheric thickness models "lith".

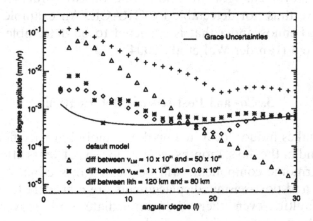

Figure 2. Comparison of the difference in degree amplitudes between three GIA models having varying viscosities and lithosphere thickness and the default GIA model, with respect to expected GRACE measurement errors (*figure taken from Wahr and Davis, 2002*).

From Figure 2 it is clear that differences in mantle viscosity and lithospheric thickness are expected to become discernible from GRACE data till about degree 15, whereby lower mantle viscosity shows the highest sensitivity. However, this figure does not show what the influence on these sensitivities is from uncertainties in present-day ice-sheet variations and from uncertainties in the Late-Pleistocene ice models.

To summarize, Figures 1 and 2 show that in the most optimistic scenarios, i.e. in the (presently about a factor of 40 too optimistic) prelaunch estimates as depicted in both figures, GRACE might be able to detect GIA motions and present-day Antarctic and Greenland ice mass decay up to harmonic degree 40, and might be able to distinguish mantle viscosity and lithosphere thickness in solid-earth models up to harmonic degree 15. This might be sufficient for discriminating between the effects of GIA above Canada and present-day Greenland ice cap changes, something which cannot be done presently with SLR dJ_n/dt observations (e.g., Vermeersen et al., 2003). Figure 2 shows that if the GRACE error curve could be lowered for harmonic degrees above 15, additional information would become available on especially lithosphere thickness and shallow mantle viscosity.

In various continental regions, seismic observations indicate the presence of shallow low-viscosity zones (intra-crustal layers, asthenosphere). Due to their shallowness, these low-viscosity zones can create high-harmonic patchlike anomalies superimposed on the low-harmonic geoid as the aforementioned broad and deep Canadian one, with typical magnitudes on the cm to m level for spatial scales of 100–1000 km (Vermeersen, 2003).

The high-harmonic geoid signatures resulting from the presence of low-viscosity zones are about one to two orders smaller than the low-harmonic geoid signatures induced by large-scale mantle flow triggered by GIA. Similarly, the temporal changes are about one or two orders of magnitude smaller and are thus not detectable by GRACE. An example of the geoid anomalies for Fennoscandia that is expected to be detectable by GOCE is given in Figure 3 (van der Wal et al., 2004).

3. Co- and Post-seismic Deformation

Large earthquakes induce local, regional and global gravity field variations, both during and in the days, months, years and tens of years after the faulting event. The harmonic components will be treated in section d; here the episodic co- and and post-seismic displacements are considered.

During a faulting event there is an immediate, non-recoverable redistribution of the Earth's mass. This is called co-seismic deformation. Due to the existence of shallow low-viscosity intra-crustal and asthenospheric layers, the redistribution of stress and strain due to the faulting will relax in the days,

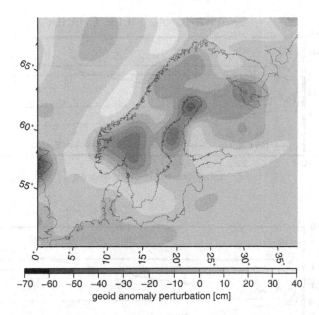

geoid anomaly perturbation [cm]

Figure 3. Simulated present-day differential geoid anomalies for the northern part of Europe due to a crustal low-viscosity zone at 20 km depth with a thickness of 12 km and a viscosity of 10^{18} Pa s, as a consequence of solid-earth deformation resulting from Late-Pleistocene Ice-Age cycles. The figure shows the geoid anomaly differences between the earth model with the aforementioned low-viscosity crustal zone and without such a low-viscosity zone in the 80 km thick lithosphere.

months, years and tens of years after the earthquake. This relaxation is not necessarily diminishing the co-seismic mass redistribution; the post-seismic deformation can enhance this.

Co- and post-seismic deformation due to large earthquakes might be detectable from space, depending on the parameters of the earthquake source, such as seismic moment (being the product of the solid-earth rigidity at the fault, fault length and relative fault displacement), type of earthquake (e.g., normal fault, strike-slip fault), geometry of the faulting event and depth of the earthquake (e.g., Sabadini and Vermeersen, 1997).

For example, Figure 4, taken from Gross and Chao (2002), shows the co-seismic effects of the great Chile earthquake of May 1960 (seismic moment of 5.5×10^{23} Nm) and the great Alaska earthquake of March 1964 (seismic moment of 7.5×10^{22} Nm), together with two largest ones during the period 1965–2000: the Sumba earthquake of August 1977, having a seismic moment of 3.6×10^{21} Nm; and the Macquarie one of May 1989, having a seismic moment of 1.4×10^{21} Nm.

Figure 4 indicates that GRACE, if it would have been active in the time frame from early 1960 to the end of 1964, should have been able to detect the co-seismic gravitational field changes of the Chile and Alaska earthquakes up to about harmonic degree 60 for the Alaskan event and up to about degree 80

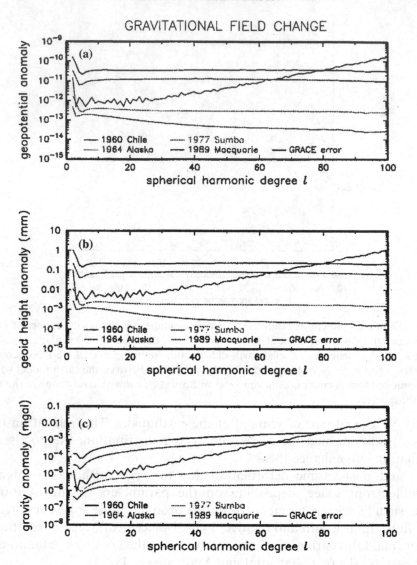

Figure 4. Gravity and geoid effects of co-seismic deformation due to four selected earthquakes, together with the expected instrumental errors of the GRACE measurements (*figure taken from Gross and Chao, 2002*).

for the Chilean event. The co-seismically induced gravitational field changes from the other two earthquakes fall below the detection level of GRACE, implying that no co-seismic effects of earthquakes would have been detected if GRACE would have been operative in the 1965–2000. Post-seismic deformation due to viscous flow of shallow low-viscosity layers in the Earth could significantly enhance or reduce the signals, and also hydrological changes associated with the stress and strain redistribution close to the fault can significantly impact crustal displacements and gravity signals.

A related issue to co-seismic deformation are ionospheric perturbations. Doppler ionospheric soundings indicate that after strong earthquakes the ionized layers E and F show displacement of several tens of meters (e.g., Artru et al., 2001). It is unclear at the moment what the gravity perturbations associated with these ionospheric layer displacements are. It might be that they are negligible in magnitude compared to the direct gravity changes from the solid Earth, but, on the other hand, the induced ionospheric perturbations are closer to the satellites than the direct solid-earth displacements.

4. Mantle Convection and Plate Tectonics

Mantle convection and one of its most prominent features, subduction of oceanic plates, show up in the geoid most conspicuously at low spherical harmonics (degrees 4–9), although also at higher degrees there are contributions (e.g., King, 2002). Apart from these quasi-static signals, it is expected that there are a number of geologically "fast" temporal changes associated with the mantle convection cylce and plate tectonics. Examples include fast rising upper-mantle plumes (e.g., Larsen, 1997), sinking slabs (e.g., Piromallo et al., 1997) and fast sinking detached slabs (e.g., Schott and Schmeling, 1998). Numerical models show that these phenomena can produce temporal geoid variation signals up to 0.1–1 mm/yr.

But higher rates are possible as well for more localized processes, e.g., for fast subsidence or emergence of oceanic islands and fast movements in volcanic regions. With the latter, also the shedding of volcanic ashes into the atmosphere during an eruptive phase might contribute to detectable temporal gravity variations.

5. Core Motions and Seismic Modes

Figure 5, taken from Crossley et al. (1999) (see also Hinderer and Crossley, 2000), gives an overview of typical normalized amplitudes of surface gravity variations due to core motions and seismic events, and the time scales on which they occur or are predicted to occur.

The "Slichter Triplet" in Figure 5 refers to a gravito-inertial translation of the solid inner core, while "FCN" and "FICN" stand for Free (Inner) Core Nutation. FCN occurs whenever there is an angle between the rotation axes of the liquid outer core and the mantle, with its eigenfrequency being proportional to the flattening of the outer core. The same with FICN, but then with respect to the solid inner core.

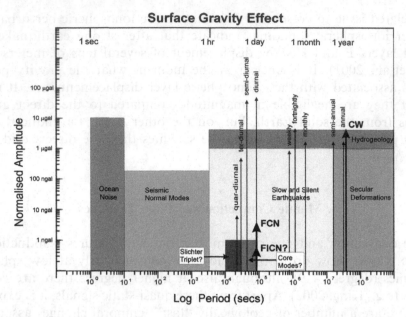

Figure 5. Typical gravimetic effects of core motions and seismic events, using harmonic amplitude harmonization (*figure taken from Crossley et al.*, 1999).

Observing the FICN would enhance our hitherto scarce knowledge of the flattening of the inner core boundary and the density jump at the inner-outer core interface.

The effects of quasi-static displacements during and after an earthquake have already been considered in the section on co- and post-seismic deformation; in Figure 5 the harmonic components of seismic faulting events are treated (*seismic normal modes*), and earthquakes that do not show up in seismograms but do show up in free oscillation observations (*slow earthquakes*) and earthquakes that do not show up in seismograms nor in free oscillation measurements (*silent earthquakes*).

Apart from the Chandler Wobble ("CW" in Figure 5), there are decadal fluctuations in the Earth's gravity field due to wobbling of the inner core. These might just reach the detection level of GRACE: Greiner-Mai et al. (2000) finds that the predicted rates of change of the Stokes coefficients C_{2m} and S_{2m} by this process averaged over a time frame of 10 years are: $C_{21} = -6.0 \times 10^{-12}$ yr^{-1}; $S_{21} = 1.0 \times 10^{-11}$ yr^{-1}; $C_{22} = -1.6 \times 10^{-12}$ yr^{-1}; and $S_{22} = -1.8 \times 10^{-12}$ yr^{-1}. The estimated standard deviations of the low-degree (<5) coefficients for GRACE are 2×10^{-12} yr^{-1} for 1 year of data, and 10^{-13} for 5 year of data. However, this decadal signal might "drown" in the contributions that other geophysical processes induce in C_{2m} and S_{2m} temporal variations.

Acknowledgements

This study was funded by Astrium under an Enabling Technology study contract. Two anonymous reviewers are thanked for their constructive comments on an earlier draft of this paper.

References

Artru, J., Lognonne, Ph., and Blanc, E.: 2001, 'Normal Modes Modelling of Post-Seismic Ionospheric Oscillations', *Geophys. Res. Lett.* **28**(4), 697–700.

Crossley, D., Hinderer, J., Casula, G., Francis, O., Shu, H.-T., Imanishi, Y., Jentzsch, G., Kaarianen, J., Merriam, J., Meurers, B., Neumeyer, J., Richter, B., Shibuya, K., Sato, T., and van Dam, T.: 1999, 'Network of Superconducting Gravimeters Benefits a Number of Disciplines', *EOS Trans. Am. Geophys. U.* **80**, 121–126.

Dziewonski, A. M. and Anderson, D. L.: 1981, 'Preliminary Reference Earth Model (PREM), *Phys. Earth Planet. Inter.* **25**, 297–356.

Greiner-Mai, H., Jochmann, H., and Barthelmes, F.: 2000, 'Influence of Possible Inner-Core Motions on the Polar Motion and the Gravity Field', *Phys. Earth Planet. Inter.* **117**, 81–93.

Hinderer, J. and Crossley, D.: 2000, 'Time Variations in Gravity and Inferences on the Earth's Structure and Dynamics', *Surv. Geophys.* **21**(1), 1–45.

Gross, R. S. and Chao, B. F.: 2002, 'The Gravitational Signature of Earthquakes', in M. G. Sideris (ed.), *Gravity, Geoid and Geodynamics 2000*, Springer, IAG Geodesy Symposia, **123**, pp. 205–210.

King, S.: 2002, 'Geoid and Topography Over Subduction Zones: The Effect of Phase Transformations', *J. Geophys. Res.* **107**(B1), doi:10.1029/2000JB000141.

Larsen, T. B. and Yuen, D. A.: 1997, 'Ultrafast Upwelling Bursting Through the Upper Mantle', *Earth Planet. Sci. Lett.* **146**, 393–400.

Piromallo, C., Spada, G., Sabadini, R., and Ricard, Y.: 1997, 'Sea-level Fluctuations Due to Subduction: The Role of Mantle Rheology', *Geophys. Res. Lett.* **24**, 1587–1590.

Sabadini, R. and Vermeersen, L. L. A.: 1997, 'Influence of Lithospheric and Mantle Layering on Global Post-Seismic Deformation', *Geophys. Res. Lett.* **24**, 2075–2078.

Schott, B. and Schmeling, H.: 1998, 'Delamination and Detachment of a Lithospheric Root', *Tectonophys.* **296**(3–4), 225–247.

Simons, M. and Hager, B. H.: 1997, 'Localization of the Gravity Field and the Signature of Glacial Rebound', *Nature* **390**, 500–504.

Tapley, B. D., Bettadpur, S., Watkins, M., and Reigber, C.: 2004, 'The Gravity Recovery and Climate Experiment: Mission Overview and Early Results', *Geophys. Res. Lett.* **31**, L09607, doi:10.1029/2004GL019920.

Tushingham, A. M. and Peltier, W. R.: 1991, 'ICE-3G: A New Global Model of Late Pleistocene Deglaciation Based Upon Geophysical Predications of Postglacial Relative Sea Level Change', *J. Geophys. Res.* **96**, 4497–4523.

van der Wal, W., Schotman, H. H. A., and Vermeersen, L. L. A.: 2004, 'Geoid Heights Due to a Crustal Low Viscosity Zone in Glacial Isostatic Adjustment Modeling: a Sensitivity Analysis for GOCE', *Geophys. Res. Lett.* **31**, L05608, doi: 10.1029/2003GL019139.

Vermeersen, L. L. A.: 2003, 'The Potential of GOCE in Constraining the Structure of the Crust and Lithosphere From Post-Glacial Rebound', *Space Sci. Rev.* **108**(1–2), 105–113.

Vermeersen, L. L. A., Schott, B., and Sabadini, R.: 2003, 'Geophysical Impact of Field Variations', in Ch. Reigber, H. Lühr, and P. Schwintzer (eds.), *First CHAMP Mission*

Results for Gravity, Magnetic and Atmospheric Studies, Springer-Verlag, Heidelberg, 165–173.

Wahr, J. M. and Davis, J. L.: 2002, 'Geodetic Constraints on Glacial Isostatic Adjustment', in J. X. Mitrovica, and L. L. A. Vermeersen (eds.), *Ice Sheets, Sea Level and the Dynamic Earth, Geodynamics Series,* American Geophysical Union, Washington, **29**, 3–32.

Earth, Moon, and Planets (2005) 94: 41–55
DOI 10.1007/s11038-005-1831-8

TIME VARIATION IN HYDROLOGY AND GRAVITY

J. RIEGGER
Institut für Wasserbau, Universität Stuttgart
(E-mail: riegger@iws.uni-stuttgart.de)

A. GÜNTNER
GeoForschungsZentrum Potsdam
(E-mail: guentner@gfz-potsdam.de)

(Received 30 September 2004; Accepted 7 February 2005)

Abstract. In view of the pivotal role that continental water storage plays in the Earth's water, energy and biogeochemical cycles, the temporal and spatial variations of water storage for large areas are presently not known with satisfactory accuracy. Estimates of the seasonal storage change vary between less than 50 mm water equivalent in areas with uniform climatic conditions to 450 mm water equivalent in tropical river basins with a strong seasonality of the climate. Due to the lack of adequate ground-based measurements of water storage changes, the evapotranspiration rate, which depends on the actual climatic and environmental conditions, is only an approximation for large areas until now, or it is based on the assumption that storage changes level out for long time periods. Furthermore, the partitioning of the water storage changes among different storage components is insufficiently known for large scales. The direct measurement of water storage changes for large areas by satellite-based gravity field measurements is thus of uttermost importance in the field of hydrology in order to close the water balance at different scales in space and time, and to validate and improve the predictive capacity of large-scale hydrological models. Due to the high spatial variability of hydrological processes temporal and spatial resolutions beyond that of GRACE are essential for a spatial differentiation in evapotranspiration and water storage partitioning.

Keywords: continental water balance, satellite gravity missions, large scale water storage, large scale evapotranspiration, sea level change, ungauged catchments

1. Introduction

The system of water redistribution within the global water cycle is the main driving force for life on the land masses. By transformation and transport processes in the hydrological cycle, water is changing its phase from liquid or ice to vapour and back. Water fluxes within and between the compartments land and ice masses, oceans and atmosphere are closely coupled to each other. In the form of a complex system of nested cycles from local up to global scales (Figure 1), mass and energy is transported over large distances. Atmospheric water vapour originating from evaporation at the ocean surface returns as precipitation on the oceans and, after vapour transport, on the

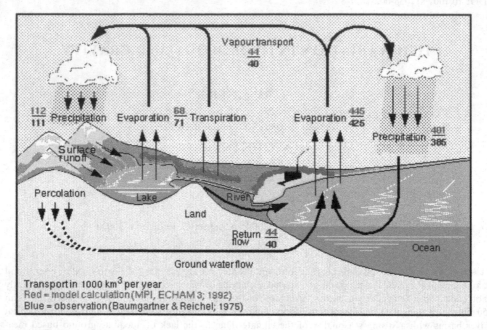

Figure 1. The global hydrological cycle (Max-Planck-Institute for Meteorology, Hamburg).

land masses. On the continents, water is recycled locally to the atmosphere by ongoing evaporation from open water surfaces or soils and by transpiration from plants and is returned by precipitation as rain or snow. These evapotranspiration processes are complex and vary considerably in time and space. They depend on the type of land use, i.e., the vegetation type, its vegetation period and leaf area, on the available soil moisture and on the local atmospheric conditions.

After withdrawal of water volumes by evapotranspiration, the remaining rain or snow melt is split up into a surface runoff component, a fast inter-flow component in the shallow soil zone and into percolation to deeper sub-surface zones resulting in a slow groundwater flow component. The relative contribution of the different flow components to total runoff is governed by topography, vegetation, soil characteristics, underlying hydrogeological conditions and the actual status of the related storages. In other words, these factors determine the relative contribution and the residence time of water masses in the different soil and rock storage compartments and, thus, the time-variable soil moisture and groundwater storage volumes. Runoff from the landscape is concentrated into the river drainage system. River runoff as well as groundwater flow at large spatial scales passes various intermediate storages, such as retention in the river network itself, in lakes or wetlands. There, it is partly subject to evaporation or extraction for human consumption, before being fed back into the oceans.

For a catchment area, being the basic spatial unit of hydrological analysis and water management issues, the water balance can be written as:

$$h_P = h_Q + h_{ET} + \Delta h_S \tag{1}$$

with h_P is the precipitation height, h_Q, the discharge height, h_{ET}, the evapotranspiration height, Δh_S, the storage change in units of water column per time interval.

However, the rates of water fluxes between the different components of the hydrological cycle vary considerably and show a specific temporal behaviour due to the different storage characteristics. These storages in the form of snow or ice cover, vegetation interception, surface water, soil moisture and groundwater all exhibit individual residence times, maximum storage levels and paths for water input and output. Characteristic average residence times of terrestrial water storages, for instance, range from a few days for the biomass or upper soil layers to several hundreds or thousands of years for deep groundwater storages (Figure 2).

Although making up only about 3.5% of total water in the hydrologic cycle, terrestrial water storage and related mass redistribution processes have a huge importance for the dynamic Earth system. Soil moisture, for instance,

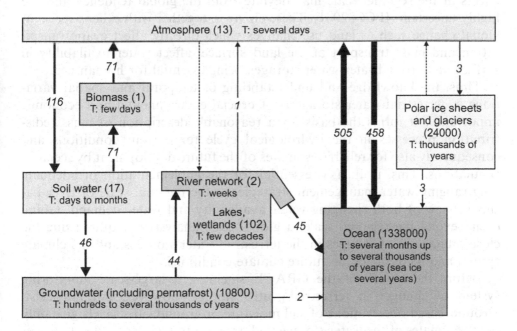

Figure 2. Storages in the global hydrological cylce. Storage volumes ($1000\,km^3$, in brackets), fluxes ($1000\,km^3$/year, in italics) and order of magnitude of mean water residence times (T) (after WBGU, 1997).

has frequently shown to be a key parameter as it links the water and energy cycles and, in addition, the biogeochemical cycle by transport of solutes and suspended load being associated with water mass redistribution.

Furthermore, terrestrial water storage is of highest importance for civilization on Earth. The replenishment of surface and groundwater storages provides the basis for water supply to a wide range of uses in the domestic, industrial and agricultural sectors. Soil moisture is essential for plant growth, including agricultural crops and thus food supply. About two-third of global water use is attributed to irrigation in agriculture. Population growth and economic development lead to an increasing water demand and rising extractions from terrestrial water storages. However, the physiographic settings of many regions in the world together with climate variability often constrain water availability to amounts being below the actual demand. About two-third of the population of the world live at least temporarily in such a condition of water stress.

Global climate change associated with a projected increase of global surface temperature in the range of 1.5–5.8 °C between 1990 and 2100 (IPCC, 2001) provides an increase of available energy for evapotranspiration and is expected to change volumes and flux rates between the storages of the hydrological cycle. Within a general tendency of increasing variability, global atmospheric water vapour and precipitation is expected to increase, although effects at the regional scale may deviate from the global tendency and are highly uncertain (IPCC, 2001). These changes, together with other aspects of global change such as land use changes which directly affect evapotranspiration and mass transport at the land surface, affect water availability in surface and groundwater water storages being essential for human use.

Thus, the knowledge and understanding of temporal and spatial variations in terrestrial water storage is of crucial environmental and economic importance. It forms the basis for a reasonable description of mass redistribution processes in the hydrological cycle for current conditions and consequently also for reliable estimates of the future development by scenario simulations. This, in turn, is essential for the implementation of adequate long-ranging water management strategies at the regional scale of river basins in view of both changing water availability and water demand. Going even beyond the regional scale, a global scale analysis is required due the close interaction of changes in the terrestrial water storages and the climate system and its feedback on future climate conditions.

Before the launch of the GRACE satellites, a large-scale monitoring system of changes in terrestrial water storage, however, did not exist. Ground-based observations of soil moisture or groundwater levels give only point estimates of the water storage and are hard to be interpolated to larger areas in view of the sparse measurement network and the multitude of influencing factors. Observations for large areas by remote sensing exist for the parameters snow cover and soil moisture, but are limited to the upper-most centimeters of the soil and do not capture the important deeper soil water and groundwater storage. While adequate measurements of precipitation and runoff may be available in some cases at the basin scale, a calculation of storage changes by use of the water balance equation (see Equation (1)) usually is not feasible as no reliable estimates of evapotranspiration are available for large scales. The shortage of adequate data (for model input and validation) also limits the applicability of hydrological simulation models to quantify water storage components for large areas.

Gravitational measurements by satellite missions are expected to be of extraordinary importance to overcome the lack of direct measurements of changes in the terrestrial water storage at large scales (e.g., Dickey et al., 1999). In the following chapters an overview is given on present open questions in hydrology and on the perspectives which are opened up by the use of gravitational measurements.

2. Global Water Balance

Until present, the global and continental water balance is not known with sufficient accuracy neither in its temporal variation nor for its mean annual values. Values differ considerably for different data sources (see Figures 1 and 3). This uncertainty is due to the difficulty of direct measurements of the climatic components of the water cycle (precipitation and evaporation) at the land surface in terms of the spatial coverage and density of measurement points. Problems also arise with the accurate measurement of river discharge on the global scale, especially for the main contributing river systems of the world with large discharge volumes. Estimates of total continental discharge into the oceans vary by about 20% (Figure 3).

Even more uncertain is the quantification of the considerably smaller net flux between oceans and land masses. It is defined by the imbalance between water vapour transport to the land masses and total runoff, and is required as a contribution to estimates of sea level change. Similar problems exist for the mass balance of ice masses (here the Antarctic and Greenland ice sheets), for which the mass output cannot be determined better than ±20% of the mass input (see Flury, 2005). However, these mass balances and the resulting net mass fluxes are essential for the determination of changes in the oceanic mass storage.

Observations of gravity changes allow a direct determination of mass variations and, thus, of net fluxes between the three compartments land masses, oceans and ice (Figure 2). However, an additional flow path has to be considered: the exchange with the atmosphere. Mass losses on land and ice masses are not only due to a loss of liquid water but also due to a release of

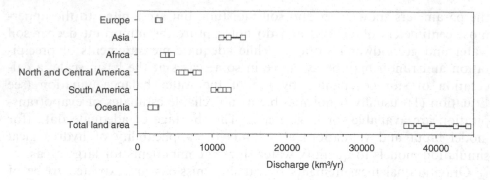

Figure 3. Long-term average annual continental discharge into oceans (estimations of six observation-based and model-based studies compared in Döll et al., 2003).

water vapour to the atmosphere. Thus, these four compartments of global water storage are closely coupled.

3. Atmospheric Mass Variations

As the gravitational measurements integrate terrestrial water and atmospheric mass variations, the effect of air mass redistributions has to be eliminated from the total signal prior to its use for hydrological analyses. The accuracy of these atmospheric corrections is of fundamental importance, as it determines the accuracy of mass variations determined for all the residual components, especially on short time scales. Local atmospheric mass per unit area is determined by the surface level pressure. In addition, the centre of gravity of each atmospheric column should be determined for signal separation. The pressure fields will usually be derived from atmospheric models such as by ECMWF (European Centre for Medium-Range Weather Forecasts) and by NCEP (National Centers for Environmental Predictions, USA) or of the respective reanalysis data. Their accuracy has to be assessed by means of a comparison with measured barometric data. It has been shown that pressure fields from operational analyses were usually adequate to remove the atmospheric contribution from GRACE gravity signals for hydrological applications with an accuracy of few millimeters of equivalent water thickness (Velicogna et al., 2001). Similarly, a pressure field derived from barometric measurements alone might be adequate if the station density is large enough. As the uncertainty in the hydrological signals due to atmospheric corrections varies with time and location, the error has to be assessed for each hydrological analysis of the gravity field measurements, depending on the area of investigation.

A direct check of the consistency of atmospheric mass changes derived from gravitational signals with that derived from surface pressure and thus a quantification of the resulting accuracy of atmospheric corrections is feasible in areas where all other fluxes causing mass changes are known or negligible. This may apply to arid zones, where after long periods without rainfall any mass redistribution by evaporation or runoff can be excluded and changes in the gravity signal are due to atmospheric mass fluxes only. Another possibility for a consistency check is to take advantage of characteristic response times of different components of the water cycle that contribute to mass variations. This may allow to separate the atmospheric signal with a high-frequency temporal behaviour from slower mass changes like groundwater storage variations.

4. Large-Scale Variations of the Terrestrial Water Storage in River Basins

Intra-annual and inter-annual dynamics of continental water storages vary substantially between environments of different physiographic and climatic conditions. For example, the intra-annual variation between maximum and minimum water storage amounts to about 50 mm of water column in river basins with rather uniform climatic conditions, whereas it is up to 450 mm in tropical river basins with a strong seasonal variation of climatic forcing, in particular precipitation input (Figure 4). These mass variations turn continental hydrology into one of the strongest signal components of time-variable gravity fields.

However, the spatial and temporal variability of water storage changes is not sufficiently known until now (e.g., singh and Frevert, 2002, for an overview). Observations of variations in continental water storages such as soil moisture or groundwater are rarely available even on small scales of sub-areas of river basins due to the limitations of the measurement methods with regard to sample density, spatial coverage or soil penetration depth as in the case of radar remote sensing of soil moisture. Yet even more difficult than to assess water storage is to quantify the water fluxes between the storages which often include complex interactions. A complementary way to quantify hydrological processes that influence water storage and fluxes is by using hydrological models. A wide range of hydrological models exists (e.g., singh and Frevert, 2002, for an overview), reaching from detailed physically-based process models to simplified models which make use of interrelated conceptual storages to represent water fluxes (e.g., evapotranspiration, percolation, runoff generation, river network routing). More comprehensive models of water management (Riegger et al., 2001) also address anthropogenic, time variant influences on water storages, such as pumping from groundwater or withdrawal from surface reservoirs for irrigation or other uses (cf. Figure 5).

The applicability of a specific model type depends, among others, on the spatial scale and the available information on soils, hydrogeology, land use

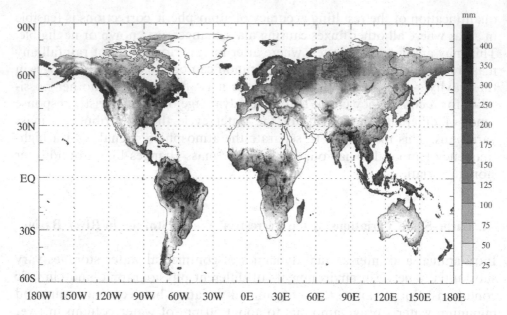

Figure 4. Average seasonal changes (changes between the months of maximum and minimum storage) of the total continental water storage (composed of the storage components snow, soil water, groundwater, river, lakes and wetlands), simulated on a 0.5° global grid with the model WGHM (Döll et al., 2003), period 1961–1995.

and climate. On small scales with detailed spatially distributed information, models can address a complex system of various interacting hydrological processes. These models often use a spatial discretization based either on a grid representation of all relevant parameters or on a sub-division of the river basin into areas of similar hydrological response.

With an increase in scale and a related decrease of detail in the available data, the actual landscape heterogeneity can no longer be explicitly represented in the model. Thus, scaling approaches are used to describe the sub-scale variability, e.g., by means of average parameters, distribution functions or simplifying lumped process formulations. In general, the capability of hydrological models to represent the hydrological cycle and, thus, their predictive power to quantify current and future variations in continental water storage, is dependent on the accuracy of input data, on the appropriateness of process formulations and on the availability of data for model calibration and validation. Large differences between regions of different climate or physiography in terms of hydrological processes and storage dynamics prevent hydrological models from being easily transferred from one region to another. In particular for large-scale applications, the only available variable for model validation usually is river discharge. Although satisfactory results may be obtained when comparing mean simulated and observed river discharge, the temporal variability and the state of soil, groundwater and surface water storage volumes may be unsatisfactorily

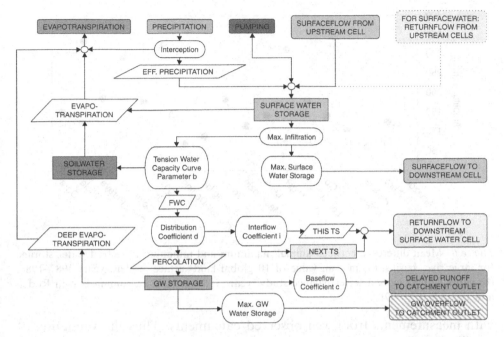

Figure 5. Flow chart of the Stuttgart–Hohai water balance model SHM (Riegger et al., 2001).

simulated in the model. As an example, current limitations of hydrological models to accurately quantify continental storage changes are shown in Figure 6 in terms of large differences in temporal storage variations between model results and water balance studies for large river basins.

In view of the existing uncertainties mentioned above, a multi-variable validation of hydrological models going beyond river discharge as validation variable has often been called for. In this respect, measurements of continental water storage changes by gravity missions can provide a unique additional data source for model validation.

First results from GRACE time-variable gravity fields reduced to the hydrological signal component clearly show a seasonal continental-scale pattern of water storage changes that corresponds to estimates by global hydrological models (Tapley et al., 2004; Wahr et al., 2004; Schmidt et al., 2005). Also at the scale of large river basins, first GRACE results allow to represent the characteristic temporal dynamics of storage change of basins in different environments. Discrepancies between water storage variations from GRACE and from hydrological models highlight, on the one hand, residual errors in the GRACE solutions (e.g., errors in reducing other than hydrological mass signals), and, on the other hand, limitations of the hydrological models.

Of course the use of satellite-based measurements as a fundamental additional information for large catchment areas has to be cross-checked

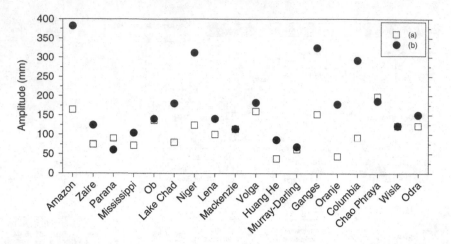

Figure 6. Mean difference between annual maximum and minimum terrestrial water storage for large river basins, (a) median value of 10 global land-surface models, years 1987–1988, (b) value derived from a water balance study, years 1989–1992 (data summarized in Rodell and Famiglietti, 1999).

with measurements from well observed catchments. Thus the validation of satellite-based gravity field measurements of continental water storage variations by ground-based measurements and the quantification of related uncertainties is of fundamental importance for hydrological modelling and forecasting and, as a consequence, for the separation of other contributions to gravity signals like the earth's mantle and crust dynamics (see Vermeersen, 2005). In principle, it consists of an investigation of the consistency between climatic and hydrological data on the one hand, and observed mass changes from satellite-based measurements on the other hand. For this comparison catchments are to be selected where ground-based measurements of soil moisture, groundwater levels, surface water storage and possibly snow cover exist with sufficient density and for which the processes are understood and reliable model estimates of water storage variations are available. An example for such a well observed and modelled catchment is the Rhine catchment (Figure 7; Hundecha and Bárdossy, 2004). With dimensions of 200×600 km and an area of $185,000$ km^2 it is at or below the limit of the spatial resolution of gravity changes measured by the GRACE mission. Rodell and Famiglietti (1999) have studied water storage changes for a number of drainage basins. They find that for basins larger than about $200,000$ km^2 reasonable changes in groundwater storage on monthly and annual time scales should be detectable by GRACE.

However, Figure 7 demonstrates that water storage variations show high temporal variability (in the order of decimeters up to 1 m of water column for this example) at much shorter spatial scales, at 100 km and less. These are certainly not detectable by GRACE. *Hydrology would benefit considerably*

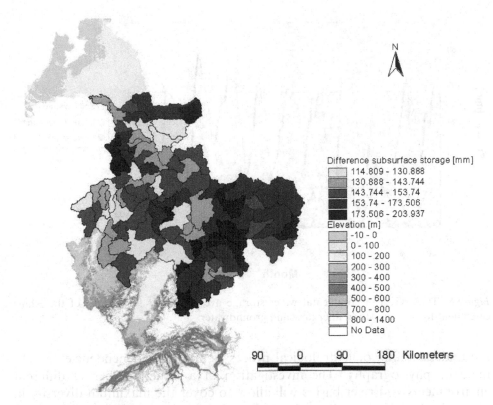

Figure 7. Spatial distribution of seasonal sub-surface water storage changes in the Rhine catchment between months of maximum and minimum storage according to the HBV model.

from future gravity missions resolving temporal variations down to small spatial scales of 100 km or less.

Gravity-based observations of terrestrial mass variations deliver integral values of storage changes of the components groundwater, soil moisture, snow water, and surface reservoir storage. Nevertheless, the characteristic temporal response of fluxes from different storages (Figure 8) – sufficient spatial resolution of 100 km or less provided – might allow a disaggregation of the gravity signal into individual components, which can be verified by ground-based measurements. This might finally allow a separation of storages even for catchments with insufficient ground-based measurements. For example, groundwater aquifers could be monitored with help of the gravity signal where a dense ground-based network does not exist, but where an increasing water demand may endanger the sustainability of groundwater resources.

The knowledge of continental water storage variations measured by satellite-based gravity field measurements is expected to considerably improve

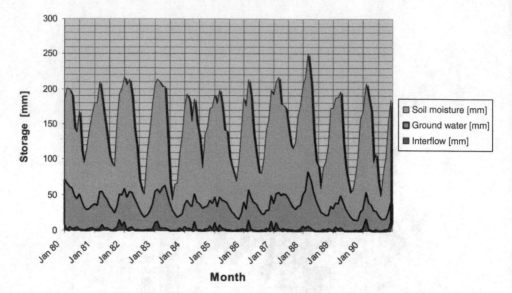

Figure 8. Time series of continental water storage averaged over 90,525 km^2 of the Rhine catchment for soil moisture, inter-flow and groundwater.

the understanding of hydrological processes and their dependencies on climate or physiography. The investigation of a large number of different environments and river basins will allow to cover the maximum diversity in basin characteristics and storage responses, ranging from humid tropical, arid and semi-arid, humid temperate to snow- and ice-dominated regions. In this way, relationships between storage variations and climate variability can be quantified for different basin characteristics and used to improve model transferability and to reduce related uncertainty. This is of particular importance for model transfer to ungauged catchments, where no calibration and validation with discharge data is possible.

5. Large-Scale Evapotranspiration

Evapotranspiration fluxes and their temporal distribution, depending on climatic conditions, actual soil moisture or the vegetation period, are poorly known on large scales. Different modelling approaches may deliver substantially dissimilar results at the monthly, seasonal or even annual time scale. Water storage changes derived from gravity variations, however, will enable to close the water balance and resolve the water balance equation for evapotranspiration (see Equation 1) (see Rodell et al., 2004, for a first application with GRACE data). This will allow for an evaluation and improvement of evapotranspiration models with respect to a realistic description under different hydrological situations (different climate zones, soil and vegetation conditions). Thus, observations of the temporal variability of continental water storages by gravity missions should be suitable for the validation of existing evapotranspiration modules in hydrological models. This reduces the degree of freedom in conceptional models and considerably enhances the quality of parameter estimations and the prognostic power of the models. The transfer of this knowledge to ungauged catchments will then allow a calculation of discharge on the basis of known water storage changes from gravity variations and climatic data and thus deliver an essential input to the assessment of the global water balance.

6. Trends and Anomalies in Continental Water Storage

Processes of environmental change may cause gradual changes in terrestrial water fluxes and storage volumes, which are of high importance for ecosystems and human water and food supply. Similarly, large-scale modifications of the oceanic and atmospheric circulation patterns such as in the course of El-Nino Southern Oscillation (ENSO) also have marked effects on the continental hydrology at inter-annual scales in terms of anomalies of precipitation, soil moisture and runoff. Inter-annual changes in the hydrological contribution to the gravity field measured by satellite missions will allow to investigate slowly changing storages, i.e. of groundwater or continental ice masses, and thus contribute to the direct detection of long-term trends for large spatial scales. New signals of climate change may become observable. In high latitudes, for instance, increasing temperature is expected to lead to a thinning or disappearance of permafrost. A long-term storage decrease will contribute to continental net discharge and thus to sea level rise. In arid to sub-humid areas, climate change is likely to decrease soil moisture, causing runoff changes, land degradation or desertification. The gravity-based observations of inter-annual water storage changes will also help to better understand the impact and persistence of global ENSO-type of circulation anomalies on continental hydrology.

Large scale anthropogenic impacts on water storage are expected to become observable by changes in the gravity field. Direct impacts by water management like withdrawal use of groundwater and surface water for irrigation as well as indirect impacts via changes in land use like deforestation or drainage of wetlands could be detected in areas where these data are not available by other means. As on large spatial scales different hydrological storages (ice, groundwater, soilwater) are coexisting, these components have to be separated by means of auxiliary data from ground measurements or hydrological models for a differentiated description of trends. Ground-based measurements of mass changes in glaciers or long-term variations in the groundwater table are therefore indispensable in order to quantify the consistency with the gravity signal and separate different storages by means of signal dynamics in well observed areas.

The dependency of the observed terrestrial storage changes on changing environmental boundary conditions in the context of climate change or human impacts is not only important for the investigation of hydrological processes and the evaluation of hydrological models, but also for the separation of geophysical contributions (by earth's mantle and crust dynamics) on this long-term time scale. In hydrologically insufficiently observed areas it is indispenible to determine long-term changes in water storage from climatic data on the basis of models. Only after the separation of hydrological gravity changes based on hydrological models, a separate analysis of geophysical components is possible.

Long-term environmental change is not only expressed by changes in the mean, but also by changes in the temporal distribution or variability. In this respect, another potential to detect gradual changes in the hydrological cycle by gravity measurements is via the analysis of changes in the intra-annual regime of storage variations on a monthly basis. Particularly, temporal shifts in the soil moisture regime due to an increasing fraction of rainfall relative to snow in the course of global warming can be analysed. Long-term changes in water storage due to changing frequencies of different atmospheric circulation patterns can potentially be detected. These analyses help to quantify the impact of environmental change, either due to natural climate variability or various anthropogenic influences, on long-term continental mass variations and changes in the hydrological cycle.

Acknowledgements

We would like to thank J. Flury, Institut für Astronomische und Physikalische Geodäsie, Technische Universität München, for reviewing and editing this contribution.

References

Alsdorf, D., Lettenmeier, D. P., and Vörösmarty, C.: 2003, 'The Need for Global, Satellite-Based Observations of Terrestrial Surface Waters', *EOS* **84**(29), 269–276.

Dickey, J. O., Bentley, C. R., Bilham, R., Carton, J. A., Eanes, R. J., Herring, T. A., Kaula, W. M., Lagerloef, G. S. E., Rojstaczer, S., Smith, W. H. F., van den Dool, H. M., Wahr, J. M., and Zuber, M. T.: 1999, 'Gravity and the Hydrosphere: New Frontier', *Hydrol. Sci. J.* **44**(3), 407–415.

Döll, P., Kaspar, F., and Lehner, B.: 2003, 'A Global Hydrological Model for Deriving Water Availability Indicators: Model Tuning and Validation', *J. Hydrol.* **270**, 105–134.

Flury, J.: 2005, 'Ice Mass Balance and Ice Dynamics from Satellite Gravity Missions', *Earth, Moon Planets*, this issue.

Hundecha, Y. and Bárdossy, A.: 2004, 'Modeling of the Effect of Landuse Changes on the Runoff Generation of a River Basin Through Parameter Regionalization of a Watershed Model', *J. Hydrol.* **292**, 281–295.

IPCC: 2001, *Intergovernmental Panel on Climate Change. 3rd Assessment report*, Cambridge University press, UK.

Riegger, J., Kobus, H., Chen, Y., Wang, J., Süß, M., and Lu, H.: 2001, 'The Water System', in P. Treuner, Z. She, and J. Ju (eds.), *Sustainable Development by Integrated Land Use Planning*, IREUS Research report 22, Institut für Raumordnung und Entwicklungsplanung, Universität Stuttgart, pp. 112–139.

Rodell, M. and Famiglietti, J. S.: 1999, 'Detectability of Variations in Continental Water Storage from Satellite Observations of the Time Dependent Gravity Field', *Water Resour. Res.* **35**(9), 2705–2723.

Rodell, M., Famiglietti, J. S., Chen, J., Seneviratne, S. I., Viterbo, P., Holl, S., and Wilson, C. R.: 2004, 'Basin Scale Estimates of Evapotranspiration Using GRACE and Other Observations', *Geophys. Res. Lett.* **31**, L20504, doi:10.1029/2004GL020873.

Schmidt, R., Schwintzer, P., Flechtner, F., Reigber, C., Güntner, A., Döll, P., Ramillien, G., Cazenave, A., Petrovic, S., Jochmann H., and Wünsch, J.: 2005, 'GRACE Observations of Changes in Continental Water Storage', *Glob. Planet. Change*, in print.

Singh, V. P. and Frevert, D. K.: 2002, *Mathematical Models of Large Watershed Hydrology*, Water Resources Publications LLC, Colorado, USA.

Tapley, B. D., Bettadpur, S. V., Ries, J. C., Thompson, P. F., and Watkins M. M.: 2004, 'GRACE Measurements of Mass Variability in the Earth System', *Science* **305**, 503–505.

Velicogna, I., Wahr, J., and Van den Dool, H.: 2001, 'Can Surface Pressure be Used to Remove Atmospheric Contributions from GRACE Data with Sufficient Accuracy to Recover Hydrological Signals?', *J. Geophys. Res.* **106**(B8), 16415–16434.

Vermeersen, B.: 2005, 'Challenges from Solid Earth Dynamics for Satellite Gravity Field Missions in the Post-GOCE Era. *Earth, Moon and Planets*, this issue.

Wahr, J., Swenson, S., Zlotnicki, V., and Velicogna, I.: 2004, 'Time-Variable Gravity from GRACE: First Results', *Geophys. Res. Lett.* **31**, L11501, doi:10.1029/2004GL019779.

WBGU: 1997, *Welt im Wandel – Wege zu einem nachhaltigen Umgang mit Süßwasser*, Wissenschaftlicher Beirat der Bundesregierung Globale Umweltveränderungen, Springer-Verlag, Berlin-Heidelberg.

Earth, Moon, and Planets (2005) 94: 57–71
DOI 10.1007/s11038-004-7606-9

FUTURE GRAVITY MISSIONS AND QUASI-STEADY OCEAN CIRCULATION

P. LEGRAND

Physical Oceanography Laboratory, Plouzane, France
(E-mail: plegrand@ifremer.fr)

(Received 8 October 2004; Accepted 14 November 2004)

Abstract. The quasi-permanent sea surface slope, i.e. the signature of oceanic currents that does not vanish when dynamic topography observations are averaged over a long period of time, will be resolved up to spatial scales of about 100 km by the GOCE space gravity mission. However, estimates of the quasi-permanent ocean dynamic topography, derived qualitatively either from models or from observations, indicate that some non-negligible residual signal remains below 100 km in areas of strong surface currents like the core of the Gulf Stream. One therefore expects that future missions can improve our knowledge of the ocean circulation in these areas. However, the potential improvements are small compared to the improvements expected from GOCE itself.

1. Introduction

Although the ocean is known to be a highly variable medium, paleoceanographic and historical records show that oceanic currents have a strong quasi-permanent signature at the surface of the ocean. Reconstructions of sea surface temperature from assemblage of planktonic foraminifera indicate that a temperature front already crossed the North Atlantic 20000 years ago, thereby suggesting the presence of a glacial equivalent of the Gulf Stream (Keffer et al., 1988; Pflaumann et al., 2003). A chart of the Gulf Stream made by Franklin more than 200 years ago puts the current at the same location as the one indicated by modem instruments (http://www.oceansonline.com/ben_franklin.htm). Of course, the instrument record averaged over a significant amount of time does not correspond to the actual current itself, but rather to the envelope within which the current occurs because of recurring latitudinal shifts and meanders over distances of several hundreds of kilometres. The important point, however, is that the signature of the currents does not vanish when the observations are averaged over a long period of time, potentially as much as several tens of thousand years, whence the concept of quasi-steady ocean surface topography. This signature shows up in altimetric measurements, which span the last 20 years only, as a departure from the geoid.

One needs to estimate the quasi-permanent ocean surface topography in order to compute the absolute velocity of surface currents from altimetric data. Until now, only low precision climatological estimates derived from in situ hydrographic measurements are available down to spatial scales of 100 km (LeGrand et al., 2003). In the near future, the GOCE gravity mission will provide a precise estimate of the marine geoid down to these scales (ESA, 1999). The question arises, however, as to whether GOCE will be able to resolve the finest spatial scales associated with the quasi-permanent ocean circulation. The issue of whether future space gravity missions will still be needed after GOCE is thus investigated here.

Section 2 begins with a quick summary of theoretical arguments that constrain the spatial scales of the ocean circulation in terms of the Rossby radius of deformation. Section 3 compares several general circulation model estimates of the spatial scales associated with the mean circulation. In Section 4, qualitative observational estimates are shown to support the model results. In Section 5, miscellaneous ocean processes that can produce intense quasi-permanent dynamic topography gradients over spatial scales not resolved by GOCE are reviewed. A special emphasis is put on regions bordering continental boundaries and on the Mediterranean Sea because of the relatively small-scale of the dynamical processes that occur there. Finally, the question of the complementary data/models that would be required to make the best use of future high resolution gravity observations is examined.

2. Spatial Scales of Oceanic Currents: Theoretical Considerations

A classical problem in geophysical fluid dynamics, first studied by Rossby (1938), is the adjustment of a surface from an initial state in which all the energy is in the form of potential energy. In this problem, an initial step

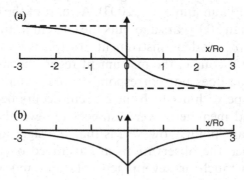

Figure 1. (a) initial step in sea surface topography (dashed line) and final topography after adjustment (full line). The horizontal scale is expressed in Rossby radii. (b) final geostrophic velocities (perpendicular to the plane).

function in sea-surface height (Figure 1a) is let to adjust under the forces of gravity and Earth rotation. The final state, obtained after the radiation of surface gravity waves, is not a state of rest but a geostrophic one (Figure 1b) in which potential energy has been converted into kinetic energy. Because of the coriolis force, the geostrophic flow is not in the direction of the pressure gradient, but at right angles, i.e. along contours of surface elevation that are parallel to the line of the initial step function. An interesting outcome of the adjustment is that the resulting geostrophic current, which has become steady after only a short time of the order of days, has a characteristic cross-stream spatial scale that is given by the Rossby deformation radius. This radius is $R_0 = (gh)^{1/2}/f$ where g is the gravitational acceleration, h is the water-depth, and f the coriolis parameter. This radius is very large, several 1000 km, and thus is not relevant to the determination of the small spatial scales of geostrophic flows.

However, the same theory applies to the adjustment of an initial perturbation in the density field in a two-layer system (Figure 2). In this case, the initial potential energy is in the form of a step function in the position of the interface between the two layers of the fluid, i.e. in the form of a density anomaly on the left-hand side of the step. This initial step will adjust in the same way as the surface step, through the radiation of internal gravity waves. After a few days, the resulting steady state current will be in geostrophic balance with a cross-stream scale also set by a Rossby radius, not the barotropic radius introduced above, but the baroclinic Rossby radius $R_1 = (g'h)^{1/2}/f$ where $g' = g\delta\rho/\rho$ is the reduced gravity, $\delta\rho$ being the density difference between the two layers (h is the thickness of the upper layer in the present case). The reduced gravity appears in the baroclinic radius because gravity no longer acts on the full density of sea water, but on the density difference between the upper layer of fluid and the lower one. Because, over most of the ocean, relatively warm surface waters overlie colder and thus denser deep waters, this two-layer baroclinic structure is characteristic of many oceanic processes. For $h = 1000$ m and $\delta\rho/\rho = 10^{-3}$ one finds a baroclinic radius of deformation on the order of 10 km. A more careful estimation carried out by Chelton et al. (1998) is presented in Figure 3. The most

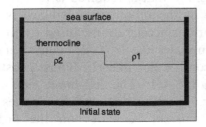 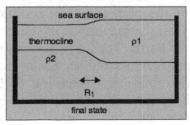

Figure 2. Initial step in the interface between two layers of fluid (the thermocline for instance) and final interface after adjustment.

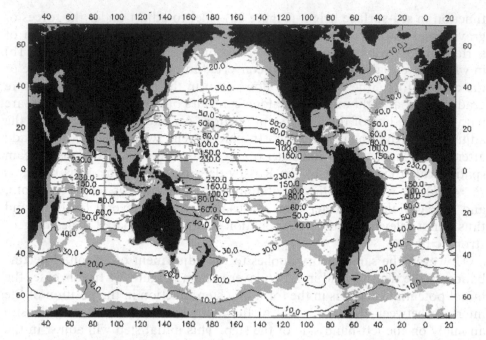

Figure 3. First baroclinic Rossby radius (km) over the world ocean (Chelton et al., 1998).

salient feature in this figure is the decrease of the radius going towards the pole caused by the increase of the Coriolis parameter. (Model results shown below are qualitatively consistent with this idea.) Thus, from theoretical arguments, one expects geostrophic currents to adjust to spatial scales beyond the reach of the 100 km resolution of the GOCE mission. Note that the adjustment of the interface between two layers of fluid has a signature at the surface of the ocean that can be detected from altimetric observations (the signal at the surface is much smaller than the signal at the interface between the two layers, unlike what is suggested in the schematic representation of Figure 2). Actually much of the ocean's variability seen by altimeters is associated with the first baroclinic mode (Wunsch, 1997).

In principle, geostrophic currents can occur at spatial scales even smaller than the first baroclinic Rossby radius. All that is required is that the magnitude of the nonlinear acceleration in the momentum balance is much smaller than the magnitude of the Coriolis acceleration caused by the rotation of the Earth. The first acceleration scales like U^2/L, U being the current speed and L its spatial scale, the second scales like fU. Thus, geostrophy remains a valid approximation if $U/(fL) \ll 1$. The left-hand side of the inequality defines the Rossby number. Therefore, geostrophy applies when the Rossby number is small. With current velocities on the order of 10 cm/s^{-1}, the geostrophic assumption is valid at mid latitudes ($f \sim 10^{-4}$ s^{-1}) as long as spatial scales of the currents are much larger than 1 km.

Geostrophic currents at these small spatial scales could for instance result from higher order baroclinic adjustments (adjustments of multiple layers of fluid). However, the associated surface signal in the open ocean is generally smaller than the signal associated with the first baroclinic mode and will probably be more difficult to observe.

3. Model Estimates of the Spatial Scales of Quasi-Steady Oceanic Currents

The characteristic spatial scales of various oceanic currents have been produced by Woodworth et al. (1998), leading to subsequent model estimates of the small spatial scales that will not be resolved by GOCE. The approach is synthesized in Le Provost and Brémond (2003). They use a 3-year average of the dynamic topography of the PAM 1/15° model to show that the amplitude of the associated small-scale dynamic topography signal can be as large as 5–10 cm in frontal zones like the Gulf Stream, the North Atlantic Current, and the East

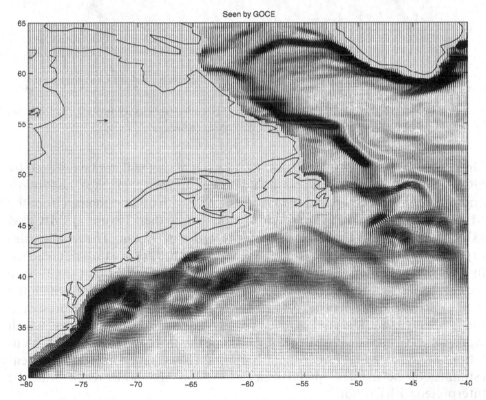

Seen by GOCE

Figure 4. Velocity signal resolved by GOCE in a 5-year mean surface velocity field produced by the 1/6° CLIPPER model. (A filter that takes the median value of the six neighbouring points has been applied to the model output to extract the spatial scales of ∼1° that will be resolved by GOCE.) Arrow in upper left corner indicates 1 m/s.

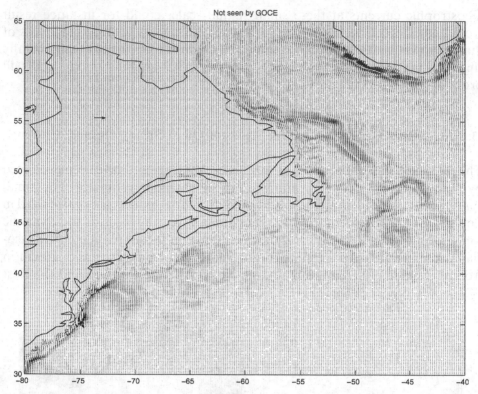

Figure 5. Velocity signal not resolved by GOCE. (The filtered velocity field of Figure 6 has been subtracted from the total 1/6° CLIPPER velocity field.)

Greenland Current (see their Figure 6). The signal that will not be resolved by GOCE is thus non-negligible and should be measurable from space. Le Provost and Bremond have compared the PAM estimate with other estimates from the MICOM and the POP high resolution models (see their Figure 3). MICOM exhibits the smallest spatial scales, on the order of 75 km. The small-scales expected from the baroclinic Rossby adjustment problem are thus not reached in the averaged model estimate. This is partly due to the fact that small-scale currents disappear in the averaged field, i.e. in the envelope within which the actual currents occur. It could also be due to the numerical diffusion present in the general circulation models, even in the very high resolution ones, which tends to smear out small-scale features. That the three different model calculations analyzed by Le Provost and Brémond (2003) give somewhat different estimates of the smallest spatial scales shows that the model results must be interpreted with caution.

In terms of current velocities, similar conclusions are reached. As an illustration, the velocity signal resolved by GOCE in another general circulation model, the 1/6° CLIPPER model (Treguier et al., 1999), is shown in

Figure 4 and the signal not resolved by GOCE is shown in Figure 5. From these figures, one sees that a gravity mission with a resolution higher than that of GOCE would allow better estimates of the small spatial scales of the Gulf Stream before it detaches from the coast, as well as better estimates of the Labrador Current. As mentioned above, the relatively large residual signal found in the latter region is consistent with the reduced Rossby radius at high latitudes. Elsewhere, i.e. in the interior of the basin, a high resolution mission would not add much to our knowledge.

Thus, numerical simulations indicate that future gravity missions will add to our knowledge of oceanic currents but the improvements, although non-negligible will be much smaller than those achieved by GOCE (LeGrand, 2001; Schroter et al., 2002).

4. Qualitative Observational Estimates of the Spatial Scales of Quasi-Steady Oceanic Currents

Because model estimates are not fully reliable, it is useful to check the model results using real data. Obviously, only qualitative checks can be made in the absence of a high resolution geoid. Two approaches are presented here that yield similar results.

A first approach is to look at sea level anomalies observed from altimetry and average them over some period of time. The idea is that the quasi-steady dynamic topography, which is the quantity of interest, has spatial scales qualitatively similar to those of averaged sea level anomalies. Indeed, dynamic topography is the sum of three terms

$$< \eta > = < \text{sla} > + \text{mssh} + g,$$

where $< \eta >$ is the dynamic topography averaged over some period of time, mssh is the mean sea surface height to which sea level anomalies (sla) are referenced, and g is the geoid. Computing mssh and sla using the first 5 years of T/P data and then, for example, averaging η and sla over the third year of T/P observations, one obtains $< \eta >$ as a function of $< \text{sla} >$ for this year. $< \text{sla} >$ can be easily computed for a particular pass of T/P, Pass 96 for instance between Greenland and Galicia (Figure 6). The one-year $< \text{sla} >$ does not contain contributions from the geoid since the sla averaged over 5 years is zero by definition and any unknown geoid contribution goes into the mssh term. Thus, it seems reasonable to assume that the spatial scales contained in $< \text{sla} >$ are qualitatively representative of the spatial scales contained in $< \eta >$. $< \eta >$ in turn is probably representative of the quasi-steady state signal since most dynamical processes acting on a one year average estimate also act on the quasi-steady circulation, and the amplitude of longer-term topography variations are relatively small. Small-scale

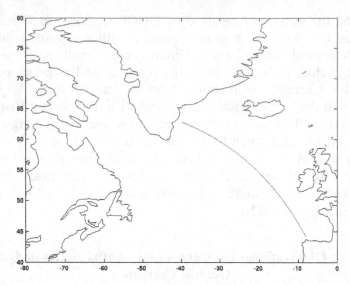

Figure 6. Location of T/P Pass 96.

features are clearly present in the estimate of $<\text{sla}>$ shown in Figure 7, for instance the 3 cm downward-slope occurring over the [1100–1150 km] segment of Pass 96. This few centimetre residual signal at a spatial scale of 50 km is consistent with the numerical model estimates, although the spatial scales found here are slightly smaller, and agree better with the Rossby radius argument.

A more complete calculation by Uchida and Imawaki (2003) combines altimetric observations of temporal variations of the sea surface height with surface drifter data to compute the absolute flow field. The resulting estimate of the mean flow field exhibits small spatial scales in the North Pacific (Figure 2a of Uchida and Imawaki, 2003) that are visually consistent with the present estimate. Unfortunately, no quantitative estimate of the signal left below 1° is provided in their paper.

A second approach is to use recent estimates of the mean wind stress curl from the QuickScat satellite radar scatterometer. Indeed, wind stress is the main driver of the ocean circulation, especially in surface layers. Moreover, it is itself affected by the underlying oceanic surface currents, so that wind stress curl patterns are attributable primarily to ocean velocities in the Gulf Stream area (Chelton et al., 2004). Figure 8 shows a map of the mean wind stress curl derived from QuickScat observations in the North Atlantic. Figure 9 shows the same map where only spatial scales smaller than 1° have been retained. This map confirms the results of the CLIPPER simulation, with some significant signal left in regions of most intense oceanic currents. The percentage amplitude of the remaining signal tends to be larger than what was found in the CLIPPER model, as much as 50% of the total signal in the

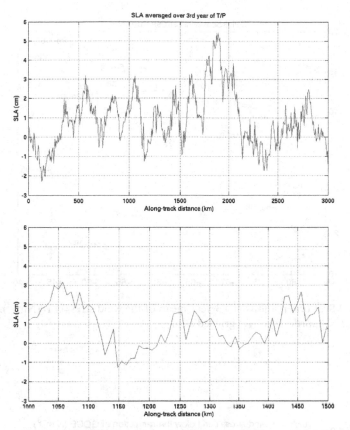

Figure 7. Upper panel: sea level anomaly between Greenland and Galicia averaged over the third year of T/P observations. Lower panel: zoom over the 1000 km–1500 km segment.

Gulf Stream before separation at Cape Hatteras. This result must be interpreted with caution, however, as the QuickScat data used here have been gridded prior to processing and only have a ½° resolution.

One therefore expects from qualitative observational estimates of the spatial scales of oceanic currents that future gravity missions could still improve estimates of the quasi-steady general circulation of the ocean after GOCE is flown. A resolution of the order of 50 km would be required with a precision better than the centimetric level. The impact on our knowledge of the quasi-steady state circulation would be significant, albeit limited.

5. Miscellaneous Particular Cases

Various processes may not be apparent in the global quasi-steady state ocean circulation, but still be of importance in local areas. A number of these processes are reviewed here, together with their implications in terms of future gravity missions.

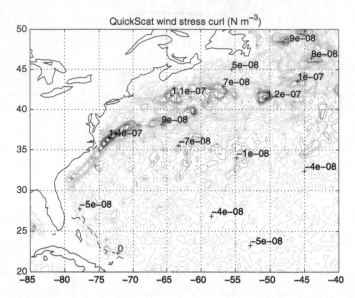

Figure 8. QuickScat observations of a 3 year-mean of the wind stress curl (Nm^{-3}). Data kindly provided by B. Chapron as a ½ ° gridded field.

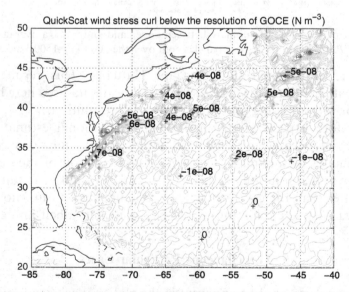

Figure 9. High-pass filtered (100 km) version of Figure 8. The filter is a simple two-dimensional median filter. The gridded observations are first averaged over neighbouring data points and are then subtracted from the original field. Qualitatively similar results are obtained using a two-dimensional Wiener filter.

5.1. BATHYMETRIC CONTROL OF THE MEAN FLOW

A class of physical processes that can cause steep mean dynamic topography gradients is the one resulting from the interaction between the flow field and the bathymetry of the bottom of the ocean. A well-known theoretical example is a jet flowing above an isolated seamount. In this example, streamlines of the flow, i.e. the dynamic topography at the surface of the ocean, tend to get closer according to the spatial scales of the underlying bathymetry (for an illustration, see Gill, 1982, p. 315). Similarly, spatial scales of flows across narrow straits (Gibraltar Strait, Florida Strait, etc.) are very strongly constrained by the underlying bathymetry. Typically, these spatial scales are a few 10s of km and will not be resolved by the GOCE mission. One therefore expects that in these areas a future high-resolution quasi-steady state mission would lead to improved estimates of the ocean circulation. Although limited in their extent, these areas are important because of their impact on the general circulation of the ocean (the spreading of Mediterranean salt tongue for instance).

5.2. NEAR-SHORE AND REGIONAL OCEANOGRAPHY

An area of great societal importance that contains intense small-scale oceanic currents is the continental boundary of the oceans. Over continental shelves, and along continental slopes, intense currents occur over spatial scales of a few 10s of km. Many of these currents being approximately in geostrophic balance, particularly the downstream component of along-slope currents (Gill, 1982, p. 378), they can be observed from space altimetry. The question, however, is whether there is any quasi-permanent signal associated with this circulation. Indeed, large temporal variations occur near-shore (including tidal variations) that dominate the signal of persistent currents. Because coastal regions are difficult to monitor because of human activities (current meter moorings for instance, are difficult to maintain near-shore because of fishing activities), little is known about the mean circulation there. The little available observational evidence hints at some relatively stable currents along the shelf break, in the Bay of Biscay for instance (B. Le Cann, personal communication). Similarly, numerical simulations exhibit a mean dynamic topography signal associated with the extension of the North Atlantic Current along the north-western coast of the UK (Figure 7 in Haines et al., 2003). It is already possible to observe these currents from satellite altimetry by computing regional estimates of the geoid through the combination of existing geoid models with in situ gravity measurements (Haines et al., 2003). Unfortunately, there are few regions where the in situ gravity data base is sufficient to undertake this approach and a remote sensing approach is required for a global coverage.

The GOCE geoid is unlikely to resolve near-shore currents because the spatial scales of the currents tend to be set by the spatial scales of the continental shelf and the continental slope, i.e. often a few 10s of km. Future very high resolution gravity missions would therefore be useful to monitor coastal regions. Were the quasi-permanent currents found to be missing, this information would in itself be interesting considering how little is known on the subject.

Another area of great societal importance is the Mediterranean Sea, which is not only bordered by very densely populated countries, but is also characterized by small baroclinic Rossby radii that are on average 11 km (Grilli and Pinardi, 1998). The spatial resolution of gravity missions would therefore matter there more than in the open ocean.

Thus, improving the knowledge of the regional and near-shore quasi-permanent ocean circulation would require very high resolutions between 10 km and 50 km and high precisions (below the centimeter level) because the mean signals, although potentially important, may be relatively weak there.

5.3. EDDY-MEAN FLOW INTERACTIONS

The study of eddy-mean flow interactions is important because of the role of eddies in redistributing momentum and heat in the ocean. In the atmosphere, where eddies have large spatial scales and are thus easier to monitor, it is known that they strongly interact with the mean flow. In the ocean, their role is less clear. Their spatial scale, which is of the order of the first baroclinic Rossby radius (Smith and Vallis, 2001), and the fact that they are ubiquitous suggest that their monitoring should be carried out from space. The problem, however, is that the mean flow being largely unobservable from altimetric data, it is not possible to evaluate precisely the mean position of fronts. Hughes and Ash (2001) show that, depending on whether the mean position of fronts in the Southern Ocean is more accurately reflected in satellite observations of sea surface temperature or in climatological in situ observations, eddies can be shown to either decelerate or accelerate the mean flow. GOCE, by providing precise and reliable observations of the mean position of fronts will improve our understanding of eddy-mean flow interactions. However, considering that eddies have spatial scales of the order of 100 km or smaller, and that topographic steering can produce very narrow jets, the resolution of GOCE may not be sufficient everywhere to accurately monitor these interactions. Future gravity missions that provide high resolution estimates of the geoid down to the Rossby radius in the Southern Ocean, i.e. a few 10s of km may thus be required, a precision at the centimeter level being probably sufficient.

6. Ancillary Data Sets

One last issue that needs to be examined is what ancillary data sets are required in order to estimate the steady circulation in combination with a high resolution quasi-steady state gravity mission, and what improvements in these data sets may be necessary.

The first complementary data set is altimetry. If one is to observe the small scales of the quasi-permanent circulation at the surface of the ocean, then altimetry must match the resolution of the gravity observations. The advent of wide-swath altimetry is very promising in this respect since this new generation of altimeters is expected to resolve spatial scales as small as 15 km while maintaining the precision level of the current generation of altimeters (Rodriguez and Pollard, 2001). Satellite altimeters of the class of Topex/ Poseidon and Jason are satisfactory in terms of precision but lack the adequate resolution in between tracks, unless several such satellites are flown simultaneously on different tracks.

Where available, in situ local gravity measurements, can greatly help in achieving the highest possible resolution and thus complement space measurements, as shown in Haines et al. (2003).

Because altimetry and gravity constrain the quasi-steady circulation at the surface of the ocean only, they need to be complemented with in situ observations in order to estimate the three-dimensional quasi-permanent flow field. No miracle is to be expected there because of the difficulty of collecting in situ measurements. The ARGO network of profiling floats (http://www.ifremer.fr/coriolis/cdc/argo.htm) already provides three-dimensional observations of temperature and salinity, but mostly in the upper 2000 m. Even this automated network leaves out big gaps in remote areas like the Southern Ocean, along the northern coast of Brazil, and in many other important regions.

Finally, precise estimates of the wind stress over the ocean, similar to those provided by the QuickScat mission, are needed in order to estimate the non-geostrophic Ekman transports in the upper layer of the ocean. It seems that the available observations are already satisfactory (see Chelton et al., 2004). However, one has to ensure that this high quality space measurement capability is maintained in the future. A solution could come from the incorporation of satellite scatterometer observations in the activities of operational agencies like EUMETSAT in Europe or NOAA in the US.

7. Conclusion

Provided that the complementary data sets enumerated above are available, particularly high resolution altimetry and scatterometer observations, one expects that some room will still be left after the GOCE mission for improvements in estimates of the quasi-permanent ocean circulation. A future very high resolution gravity mission resolving spatial scales of about 50 km would be required. It seems reasonable to expect that improved estimates of these scales of the ocean circulation could locally lead to corresponding improvements in estimates of heat transport. However, obtaining observations of the three-dimensional temperature field on a 50 km grid would be very difficult in the global ocean.

The improvements provided by a future high spatial resolution mission would by no mean match those expected from the GOCE mission. However, processes like intense oceanic fronts, shelf-break currents, and flows in semi-enclosed seas could be determined more accurately. Most of these processes have small geographical extents, but they can have a strong impact on the global circulation and are of great societal importance.

References

Chelton, D. B., deSzoeke, R. A., Schlax, M. G., El Naggar, K., and Siwertz, N.: 1998, 'Geographical Variability of the First-Baroclinic Rossby Radius of Deformation', *J. Phys. Oceanogr.* **28**, 433–460.

Chelton, D. B., Schlax, M. G., Freilich, M. H., and Milliff, R. F.: 2004, 'Satellite Measurements Reveal Persistent Small-Scale Features in Ocean Winds', *Science* **303**, 978–983.

European Space Agency: 1999, 'The Four Candidate Earth Explorer Core Missions – Gravity-Field and Steady-State Ocean Circulation', ESA report SP-1233, ESA Publication Division, c/o ESTEC, Noordwijk, The Netherlands.

Gill, A.E.: 1982, *Atmosphere-Ocean Dynamics*, Academic Press, New York, 662 pp.

Grilli, F. and Pinardi, N.: 1998, 'The Computation of Rossby Radii of Deformation for the Mediterranean Sea', *MTP news* **6**, 4.

Haines, K., Hipkin, R., Beggan, C., Bingley, R., Hernandez, F., Holt, J., Baker, T., and Bingham R. J.: 2003, 'Combined Use of Altimetry and in situ Gravity Data for Coastal Dynamics Studies', *Space Sci. Rev.* **108**, 205–216.

Keffer, T., Martinson, D. G., and Corliss B. H.: 1988, 'The position of the Gulf Stream during quaternary deglaciations', *Sci.* **241**, 440–442.

LeGrand, P.: 2001, 'Impact of the Gravity Field and Steady-State Ocean Circulation Explorer (GOCE) Mission on Ocean Circulation Estimates. Volume Fluxes in a Climatological Inverse Model of the Atlantic', *J. Geophys. Res.* **106**, 19,597–19,610.

LeGrand, P., Schrama, E. J. O., and Tournadre J.: 2003, 'An Inverse Modeling Estimate of the Dynamic Topography of the Ocean', *Geophys. Res. Lett.* **30**, 1062–1065.

Le Provost, C. and Brémond M.: 2003, 'Resolution Needed for an Adequate Determination of the Mean Ocean Circulation from Altimetry and an Improved Geoid', *Space Sci. Rev.* **108**, 163–178.

Pflaumann, U. et al.: 2003, 'Glacial North Atlantic: Sea Surface Conditions Reconstructed by GLAMAP 2000', *Paleoceanography* **18**(3), 1065, 10-1–10-28.

Rodriguez, E. and Pollard B. P.: 2001, 'Report of the High-Resolution Ocean Topography Science Working Group Meeting', D. B. Chelton (ed.), *College of Oceanic and Atmospheric Sciences*, OSU, Corvallis, Oregon.

Rossby, C. G.: 1938, 'On the Mutual Adjustment of Pressure and Velocity Distributions in Certain Simple Current Systems'. II, *J. Mar. Res.* **2**, 239–263.

Schroter, J., Losch, M., and Sloyan B.: 2002, 'Impact of the Gravity Field and Steady-State Ocean Circulation Explorer (GOCE) Mission on Ocean Circulation Estimates. Volume and Heat Transports Across Hydrographic Sections of Unequally Spaced Stations', *J.Geophys.Res.* **107**, 4-1–4-20.

Smith, K. S. and Vallis, G. K.: 2001, 'The Scales and Equilibration of Midocean Eddies: Freely Evolving Flow', *J. Phys. Oceanogr.* **31**, 554–571.

Treguier, A. M., Reynaud, T., Pichevin, T., Barnier, B., Molines, J. M., de Miranda, A. P., Messager, C., Beismann, J. O., Madec, G., Grima, N., Imbard M., and Levy C.: 1999, 'The CLIPPER Project : High Resolution Modeling of the Atlantic', *Int. WOCE Newslett.* **36**, 3–5.

Uchida, H. and Imawaki, S.: 2003, 'Eulerian Mean Surface Velocity Field Derived by Combining Drifter and Satellite Altimeter Data', *Geophys. Res. Lett.* **30**(5), 1229, doi:10.1029/2002GL016445.

Woodworth, P. L., Johannessen, J., LeGrand, P., Le Provost, C., Balmino, G., Rummel, R., Sabadini, R., Sunkel, H., Tscherning, C. C., and Visser P.: 1998, 'Towards the Definitive Space Gravity Mission', *Int. WOCE Newslett.* **33**, 37–40.

Wunsch, C.: 1997, 'The Vertical Partition of Oceanic Horizontal Kinetic Energy', *J. Phys. Oceanogr.* **27**, 1770–1794.

Earth, Moon, and Planets (2005) 94: 73–81
DOI 10.1007/s11038-005-0452-6

FUTURE BENEFITS OF TIME-VARYING GRAVITY MISSIONS TO OCEAN CIRCULATION STUDIES

CHRIS W. HUGHES

Proudman Oceanographic Laboratory, 6 Brownlow Street, Liverpool, L3 5DA, UK
(E-mail: cwh@pol.ac.uk)

PASCAL LEGRAND

Physical Oceanography Laboratory, IFREMER Brest, BP 70, 29280, Plouzane, France
(E-mail: pascal.le.grand@ifremer.fr)

(Received 9 August 2004; Accepted 3 June 2005)

Abstract. A summary is offered of the potential benefits of future measurements of temporal variations in gravity for the understanding of ocean dynamics. Two types of process, and corresponding amplitudes are discussed: ocean basin scale pressure changes, with a corresponding amplitude of order 1 cm of water, or 1 mm of geoid height, and changes in along-slope pressure gradient, at cross-slope length scales corresponding to topographic slopes, with a corresponding amplitude of order 1 mm of water, or a maximum of about 0.01 mm of geoid. The former is feasible with current technology and would provide unprecedented information about abyssal ocean dynamics associated with heat transport and climate. The latter would be a considerable challenge to any foreseeable technology, but would provide an exceptionally clear, quantitative window on the dynamics of abyssal ocean currents, and strong constraints on ocean models. Both options would be limited by the aliassing effect of rapid mass movements in the earth system, and it is recommended that any future mission take this error source explicitly into account at the design stage. For basin-scale oceanography this might involve a higher orbit than GRACE or GOCE, and the advantages of exact-repeat orbits and multiple missions should be considered.

Keywords: Abyssal, ocean circulation changes, ocean bottom pressure, satellite gravity

1. Introduction

Time-varying gravity is, over the ocean, potentially a measure of ocean bottom pressure (OBP) variations. Such variations have been studied from *in situ* data predominantly at short (days to months) time scales and at occasional points in the global ocean. The possibility of obtaining multi-year time series of global OBP variations is a very exciting one, and one which is still quite a new concept to the oceanographic community. The first results from GRACE are now beginning to demonstrate this possibility. The future launch of GOCE, which will refine knowledge of the static geoid at smaller length scales, will not significantly address the question of time-dependent gravity relevant to OBP measurement. The purpose of this paper is to look at the possible OBP signals to be measured, for consideration in the design of future systems for measuring time-dependent gravity.

There are two distinct length scales on which OBP variations can be considered: "basin scales" of typically 1000–4000 km, and topographic length scales of about 20–200 km. In what follows, the dynamical information associated with these two length scales is discussed separately.

Currently, however, the theoretical limiting factor on accuracy of OBP determinations obtainable from a GRACE-like mission appears to be the need to correct for high frequency mass movements which are too fast to be measured with the global coverage that is necessary to map their spatial signatures and subtract them from long-period variations. Some thoughts on the requirements of future missions to minimise this problem will also be offered. It is worth noting that, at the time of writing, the accuracy of GRACE data is at least an order of magnitude worse than its theoretical value. No clear reason for this has yet been identified, as the instruments all appear to be performing close to or better than expectations. It is anticipated that significant improvements can be expected as the data processing chain is refined.

2. Basin Scale Variability

On basin scales, it has long been predicted (Gill and Niiler, 1973), and more recently observationally verified (Fukumori et al., 1998), that most of the intraseasonal to seasonal OBP variability is due to the ocean's depth-independant response to atmospheric wind stress and pressure forcing. This is of little intrinsic interest except as a source of aliassing for altimetry and temporal satellite gravity measurements, which cannot resample the same spot frequently enough to remove this, sometimes large (up to 5 cm of water in the deep ocean), high frequency signal from longer term measurements.

At interannual scales, however, the dynamics become much more interesting. The global heat budget is strongly influenced by the ocean circulation, which supplies almost half of the equator-to-pole heat transport which ameliorates what would otherwise be an enormous temperature range around the planet. In the Atlantic, this heat transport is actually northwards in both hemispheres. Such large heat transports are made possible by the sinking of cold water at high latitudes, and spreading throughout the ocean at depth. There are rather few sites of bottom water formation, in the North Atlantic and close to Antarctica, and this water spreads over the whole globe. From the few direct measurements we have, and from the spreading of tracers advected by the slow, deep flows, we know something about the spreading pathways the water takes. It is also known that there are strong variations in production of dense bottom waters as this tends to happen during extreme weather events and is quite sensitive to ambient ocean conditions. However, very little is known of the variability in the flow once it sinks to the ocean floor, or of variations in the subsequent upwelling. This is because direct measurements of deep flows are difficult to make, and therefore rather sparse. In some cases, deep flows may be detectable indirectly by their influence on near-surface circulations, but this is an indirect representation of the deep flow and may be swamped by other near-surface dynamics, making it difficult to infer the deep flow.

Movement of dense water between ocean basins must, however, produce a signal in ocean bottom pressure. A simple analytical argument (Hughes and Stepanov, 2003) shows that this signal will be predominantly at basin scales as ocean dynamics will rapidly smooth the pressure along depth contours. There is a strong dynamical constraint on the component of pressure gradient along a depth contour, but no such constraint on the pressure averaged around a closed depth contour. This permits a range of possible "basin scale" pressure signals: signals may be basin scale along a depth contour, but at a relatively narrow scale across depth contours (for short, closed depth contours, such as those around a seamount, both scales can be small), or can be basin scale in both directions. The latter are susceptible to measurement using current capability.

Multi-year time series of bottom pressure data at basin scales, in combination with more routine near-surface measurements, will make it possible to track the motion of dense water through the global ocean, from its source to its still poorly-defined upwelling regions, making a major contribution to monitoring and understanding the ocean-atmosphere climate system. Such a monitoring system would ideally be a part of an ongoing monitoring effort to measure interannual to decadal changes in ocean circulation.

It is difficult to specify the size of the expected signals in advance. Modelling results (Hughes and Stepanov, 2003) suggest typical interannual pressure changes of around 1 cm of water (equating to about 1 mm change in geoid at these length scales), but there is now strong evidence that much larger changes can occur. Using Lageos satellite tracking data, Cox and Chao (2002) presented a time series of the J_2 coefficient of the earth's gravity field. This shows a steady decrease, attributable to the secular relaxation of the earth after melting of ice from the last glacial period, but also shows a sudden jump upwards (of about 10^{-10}) in mid 1998. Dickey et al. (2002) attributed about half of this jump to a change in OBP, predominantly in the Southern Ocean. While this attribution depends heavily on data assimilation into a coarse resolution ocean model (which is unlikely to represent the strongly eddying Southern Ocean well, and is limited by a dearth of subsurface data in this region), it is undoubtedly true that there was a major sea level drop of about 15 cm in the Pacific sector of the Southern Ocean at this time

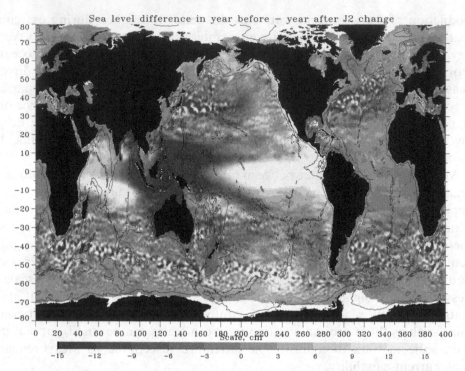

Sea level difference in year before − year after J2 change

Scale, cm

Figure 1. The difference between two annual average sea level fields, year before minus year after the change in J_2, from satellite altimetry. The 3 km depth contour is superimposed.

(Figure 1). If this is associated with a mass change, the associated geoid movement is of order 1 cm.

This drop occurred at the same time as the switch from El Niño to La Niña conditions, as can be seen from the large tropical Pacific sea level changes, and highlights both the possible interest of satellite gravity measurements and also the difficulty in interpreting other forms of data. The sea level change in the Southern Ocean is likely to be associated with a mass anomaly, since Southern Ocean flows tend to penetrate to the ocean floor, although model diagnostics suggest that the relationship between sea level and OBP is geographically surprisingly variable in the Southern Ocean (R. Bingham, personal communication). In the tropics, however, a sea level rise tends to be strongly affected by heating, and compensated by a depression in the thermocline, producing a much smaller bottom pressure signal. This means that sea level is a poor guide to movements of mass at low latitudes, and cannot tell us where the "missing mass" removed from the Southern Ocean has gone. Interpretation of model data is also fraught with difficulties given the small amount of data from the depths of the ocean to constrain the models. It is likely that the sea level change is simply the surface signature of a much larger global redistribution of mass which we cannot see.

Even in itself, this 15 cm sea level change is enough to account for 1/3 to 1/2 of the change in J_2, if we assume the mass removed from the Southern Ocean is redistributed at tropical latitudes.

This is a completely unexpected phenomenon. It is unclear whether its relationship to El Niño is coincidence or something deeper, but it represents a major shift in the ocean in this region. Interpretation would be greatly helped if the mass changes could be more localised, as by a GRACE-type mission. Such a major discovery, associated with the most climatically important and most thoroughly studied ocean phenomenon, shows how interesting even one coefficient of the gravity field can be.

Of equal topical interest is the possibility that the North Atlantic might switch into a different mode of circulation in which much less bottom water is formed, with possibly dramatic consequences for the Northern Hemisphere. Palaeoceanographers argue that such changes occurred during the glacial-interglacial transition (Clark et al., 2002), and there is concern that current anthropogenic climate forcing could tip the system into a state where such a "thermohaline collapse" could occur again. There are several international programmes as part of CLIVAR, whose objective is to monitor the ocean for such changes, but their geographical range is necessarily limited; temporal gravity measurements would be a great help in such monitoring efforts.

The projected accuracy of the current GRACE mission would be easily adequate to resolve these kinds of variability, with an estimated accuracy of 1 mm of water for a length scale of about 1000 km (Wahr et al., 1998), although the present actual accuracy appears to be closer to 1 cm of water at basin scales. However, that is a measure of the instrument accuracy, ignoring the considerable problem of aliassing high frequency variability. If the processing of GRACE data can be improved to boost the system accuracy, the problem of unmodelled tides and barotropic ocean variability particularly in shelf seas and at high latitudes, is likely to become the dominant source of noise (Knudsen, 2003; Ray et al., 2003; Hughes and Stepanov, 2003).

3. Topographic Scale Variability

In order to go further than monitoring basin-scale mass redistributions, a leap in resolution is necessary to permit more detailed attribution of the associated dynamics. It will never be possible to use gravity alone to monitor time-averaged flows, but bottom pressure changes can contain information about changes in the currents.

Currents associated with these deep mass transports are typically concentrated against steep topography, with length scales between about 20 and 200 km. A simple estimate of the gravity anomaly associated with subtraction of a uniform 0.2 kg/m^3 density anomaly, occurring over a 1 km thick layer and 100 km horizontal extent, produces a geoid anomaly of 1.2 mm. Removal of such a density anomaly is equivalent to a change of about 20 cm of water, and would represent the complete extinguishing of a major deep western boundary current, as discussed above in the case of the North Atlantic. This is an extreme upper bound to the size of signal which can be expected, we would expect more usual changes to be at least an order of magnitude smaller. Thus, while movement of dense water between basins can lead to a basin-wide signal of a few cm of water, the fluctuating currents which lead to these transports are associated with changes of similar magnitude, but over smaller length scales, making them correspondingly harder to detect from space.

Going further than this, we can consider the possibility of measuring the much smaller variation of pressure along depth contours. Hughes and de Cuevas (2001) demonstrated the importance of the interaction between bottom pressure and topography for the depth-integrated ocean flow. In principle, if wind stress, topography, and OBP were perfectly known, the depth-integrated flow could be deduced on length scales of about 200 km. As noted before, the absolute value of OBP will never be determined from gravity alone, but fluctuations could in principle be related to fluctuations of the depth-integrated flow, given a reasonable knowledge of the topography. The difficulty lies in the small signals, and small length scales needed: the dynamically significant variable is the gradient of pressure along a depth contour, so pressure needs to be known on the length scale over which topography varies, which can be as small as 10 km. In addition, the strong constraint resulting from this relationship means the associated pressure signals are of order a few millimeters of water, even for quite large changes of transport (although the pressure change across depth contours can be much larger). For a 1 mm change at 50 km length scale, this would require a geoid accuracy a factor of 400 better than that considered above for a 20 cm change at 100 km resulting in the need to measure the geoid to an accuracy of at least 0.03 mm on length scales of 50 km, or higher for smaller length scales. Benefits to ocean circulation studies are potentially very large, but the technical challenge is also very large.

4. Accuracy Requirements

As stated in Section 2, The projected accuracy of GRACE is sufficient to measure basin scale variations, although current error estimates suggest that in practice this capability is marginal. Should the projected accuracy be achieved, the problem would then be with aliassing of high frequency variability. While every effort is being made to model as much of this as possible, it is likely to remain the major error source for the forseeable future. Global barotropic models will certainly develop, and may make a significant impact on storm surge modelling for shelf seas if they reach adequate resolution (and if adequate meteorological forcing data are available). Tides are still a problem, particularly in the Arctic (Ray, 2003), and there are few observations against which to test models.

A good argument can be made for a gravity mission specifically designed to address these problems. There is an analogy in satellite altimetry, in which Geosat demonstrated the potential, but TOPEX/Poseidon first really unlocked that potential by choosing an orbit specifically designed to sample the tides efficiently, thus permitting them to be accurately removed from the signal. Even the relatively straightforward sampling of the ocean by Geosat produced important ambiguities between tidal error signals and long period ocean dynamics. If we think of GRACE as analogous to Geosat, we can use the information gathered by GRACE to help design a mission to minimise such ambiguities.

If basin scale dynamics are the aim, it is possible that the optimum configuration would be an orbit higher than the GRACE orbit, enhancing the spatial averaging capability of each satellite pass and improving the temporal sampling. The spatial resolution would go down in a "no high frequency noise" scenario, but in the real world the useful spatial resolution may improve. An exact repeat orbit would make interpretation of aliasing much simpler, and mutiple instruments could be used to improve the trade-off between temporal and spatial sampling. In addition, if the orbit is designed with sampling of tides in mind, a major source of error could be rapidly reduced.

For topography scale dynamics (say 100 km), the requirements are much stricter. The maximum plausible signal is about 1 mm of geoid, which is beyond the capability of both GOCE and GRACE. However, the principle interest is in ocean signals at long time scales. If technological improvements could improve the GOCE precision from 1 to 0.5 cm over 6 months, then simple statistical averaging could make 1 mm a resonable aim for the difference of two 10 year averages. Further improvements in technology (such as superconducting gradiometers and laser tracked satellite constellations) may make fractions of a millimetre a plausible aim, in which case the improved spatial resolution of a low orbit, gradiometry mission may help with the aliassing problem by using the different spatial distributions of the main high frequency "noise" terms from the lower frequency "signal" terms. The problem of resolving topographically-steered boundary current changes is made more difficult by their usual proximity to land, meaning that leakage of land hydrography signals (which can be much larger) will also be a problem. In short, resolution of topography-scale dynamics is a big challenge which may just barely be attainable with a major improvement in accuracy.

To go further and diagnose transport fluctuations directly from changes in the gradient of pressure along depth contours would require a further order of magnitude improvement in both measurement accuracy and reduction of aliasing error. Although the dynamical value of such measurements would be high, this does not seem a plausible avenue to explore for the foreseeable future.

Whichever option is followed, it seems clear that the mission design is critical. If signal and noise are to be separated, a judicious choice of orbit(s), duration, and measured parameters must be made which takes account not only of instrument noise but also geophysical noise to optimise detection of the desired signal. This would be different for different signals. For example, basin-scale ocean dynamics might best be detected from a slightly higher orbit which would reduce temporal aliasing, but land hydrology has a larger signal and may have a smaller aliasing problem, and might benefit from the lowest orbit possible. As has been found with satellite altimetry, the best exploitation of the technology, minimising aliasing problems, may require multiple gravity missions (and perhaps a mixture of technologies) in order to optimise both the spatial and temporal sampling.

5. Conclusion

Important new understanding of how the ocean works will undoubtedly be derived from measurements of time-dependent gravity, as can be seen from the interest already generated by changes in a single coefficient (J_2). The easier aim, which will provide a first global window on ocean thermohaline circulation variations and their role in heat transport, is to measure basin scales of 1000–4000 km, but with an instrument designed to minimise the impact of geophysical noise. Geoid signals of order 1 mm or more are to be expected, within the range of current technology. Moderate accuracy improvements would increase the spatial resolution attainable and aid the interpretation of the signal, but would not produce a step change in value of the observing system.

The more ambitious aim is to look for resolutions in the range of 20–200 km capable of collocating pressure anomalies with topographic features. The most gross changes in circulation would produce signals of order 1 mm in the geoid, while more subtle signals containing important dynamical information would require accuracies of order 0.01 mm on length scales smaller than 100 km. Should that be achievable, however, the gain for oceanography would be enormous, permitting the depth-integrated flow to be calculated at about 200 km resolution and giving a clear picture of flows near the ocean floor. The current and projected observational network is only capable of providing very limited information on such flows. However, the accuracy needed for the more subtle signal of pressure gradient along topography appears beyond anything currently foreseeable.

Whatever the aim of any future mission, a serious complicating factor which must be take into account is aliasing of high frequency signals. Lessons must be learned from GRACE to help design a mission so as to minimise these errors. The trade-off between temporal and spatial resolution should be carefully examined in the light of the signal to be measured, as should the potential advantages of an exact repeat orbit. The clear interpretation of both ocean and land signals may require multiple missions.

References

Clark, P. U., Pisias, N. G., Stocker, T. F., and Weaver, A. J.: 2002, *Nature* **415**(6874), 863–869.

Cox, C. M. and Chao, B. F.: 2002, *Science* **297**, 831–883.

Dickey, J. O., Marcus, S. L., de Viron O, O., and Fukumori, I.: 2002, *Science* **298**, 1975–1977.

Fukumori, I, Raghunath, R., and Fu, L.-L.: 1998, *J. Geophys. Res.* **103**, 5493–5512.

Gill, A. E. and Niiler, P. P.: 1973, *Deep-Sea Res.* **20**, 141–177.

Hughes, C. W. and de Cuevas, B. A.: 2001, *J. Phys. Oceanogr.* **31**, 2871–2885.

Hughes, C. W. and Stepanov, V.: 2003, *Space Science Reviews* **108**, 217–224.

Knudsen, P.: 2003, *Space Science Reviews* **108**, 261–270.

Ray, R. D., Rowlands, D. D., and Egbert, G. D.: 2003, *Space Science Reviews* **108**, 271–282.

Wahr, J., Molenaar, M., and Bryan, F.: 1998, *J. Geophys. Res.* **103**(B12), 30205–30229.

Earth, Moon, and Planets (2005) 94: 83–91
DOI 10.1007/s11038-004-8213-5

ICE MASS BALANCE AND ICE DYNAMICS FROM SATELLITE GRAVITY MISSIONS

J. FLURY

Institut für Astronomische und Physikalische Geodäsie, Technische Universität München
(E-mail: flury@bv.tum.de)

(Received 17 August 2004; Accepted 27 December 2004)

Abstract. An overview of advances in ice research which can be expected from future satellite gravity missions is given. We compare present and expected future accuracies of the ice mass balance of Antarctica which might be constrained to 0.1–0.3 mm/year of sea level equivalent by satellite gravity data. A key issue for the understanding of ice mass balance is the separation of secular and interannual variations. For this aim, one would strongly benefit from longer uninterrupted time series of gravity field variations (10 years or more). An accuracy of 0.01 mm/year for geoid time variability with a spatial resolution of 100 km would improve the separability of ice mass balance from mass change due to glacial isostatic adjustment and enable the determination of regional variations in ice mass balance within the ice sheets. Thereby the determination of ice compaction is critical for the exploitation of such high accuracy data. A further benefit of improved gravity field models from future satellite missions would be the improvement of the height reference in the polar areas, which is important for the study of coastal ice processes. Sea ice thickness determination and modelling of ice bottom topography could be improved as well.

Keywords: Geoid time variation, ice mass balance, ice thickness, satellite gravity missions, sea level change

1. Introduction

Understanding the mechanisms controlling ice sheet mass balance is essential to studies of long term changes in sea level and ocean circulation and of climate change. The masses of the Antarctica and Greenland ice sheets, of glaciers and ice caps vary in time through the exchange of water with the atmosphere and with the oceans, see Committee (1997). Ice mass balance is the comparison of mass gain due to accumulation and mass loss due to sublimation, evaporation, melting and discharge, as functions of time, respectively. In the near future detailed mass balance will be feasible from the combination of models (atmospheric models, flow models, surface energy balance), *in-situ* measurements (ice cores, surface GPS, absolute gravimetry, automatic weather stations), airborne measurements (shallow layer radar, ice penetrating radar, and airborne gravimetry) and spaceborne measurements (satellite microwave, INSAR, satellite radar and laser altimetry, and satellite gravity), see Figure 1 (from Thomas, 2001). Temporal variations of gravity are of particular importance as they are directly proportional to mass changes. The challenge will be to separate postglacial isostatic mass adjustment, the effect of ice compaction and the mass changes of ice, see Wahr and Velicogna (2003).

2. Ice Mass Balance (Ice Sheets and Glaciers)

The history of past glaciation and the present-day ice mass balance are parts of a complex process: past ice load changes, in particular the deglaciation after the last ice age, continue to act through Glacial Isostatic Adjustment (GIA), i.e. vertical land movements due to removed ice loads and related lateral mass shifts in the Earths's interior. The GIA geometric and mass signals are superimposed by the recent ice thickness and mass changes, caused by the meteorological and climatological variability, with seasonal and annual to interannual and secular components. Measurements of geometry and mass changes in time always represent the total of these effects, and – in addition – of other phenomena such as variations in the atmosphere, in adjacent oceans and – for areas free of ice today – also in continental hydrology (cf. Vermeersen, 2004). Ice mass and sea level are closely connected by the hydrological cycle including ice, ocean and atmosphere.

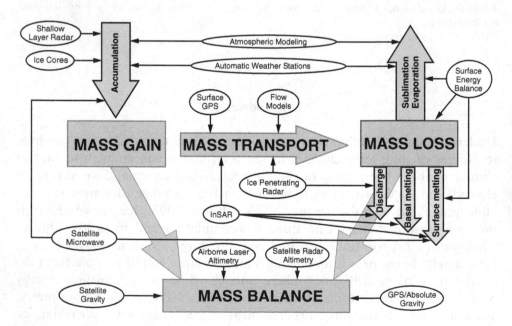

Figure 1. Processes contributing to ice mass balance and applied measurement methods, after Thomas, 2001).

Therefore, ice mass balance is often expressed in units of equivalent sea level change.

Figure 1 shows the wide variety of data connected to ice mass changes: ice cores and geological records revealing past climate changes (Thomas, 2001), relative sea level changes from tide gauges (Woodworth, 2004), meteorological data and models, surface ice velocities from INSAR, surface structure and temperature from satellite microwave data, vertical motion from GPS, changes in the low harmonic coefficients of the gravity field and in the earth orientation parameters (James and Ivins, 1997), gravity changes as measured by terrestrial absolute gravimetry, and others.

All these data have characteristic limitations in their role for the determination of the ice mass balance. Often the sampling is sparse and uneven, resulting in a limited spatial and temporal resolution. For others the conversion of the measured quantity into mass variations contains many error sources.

From the currently available data it has been deduced, that the Antarctica ice sheet is today approximately in equilibrium. However, the uncertainties of this conclusion are large. Table I (first row) shows the current accuracy estimates as given by various authors, ranging from ± 0.2 mm/year up to ± 1.4 mm/year for the resulting secular change in sea level equivalent. In the mass budget, the mass output is constrained only to about $\pm 20\%$ of the mass input (Huybrechts et al., 2004). The seasonal and interannual components are probably even more uncertain. Also for Greenland, the mass balance is poorly known. Here, melting and discharge prevail, resulting in a sea level rise of at least 0.1 mm/year, maybe considerably more (Rignot and Thomas, 2002).

TABLE I

Ice mass balance of the Antarctica ice sheet: current and future accuracies for secular ice mass change and corresponding sea level equivalent, as estimated for the spatially averaged change of the whole ice sheet

	Accuracy estimates (mm/year)		References
	Secular ice mass change	Secular sea level change	
today	20–40	0.2–1.4	James and Ivins (1997); Wahr et al. (2000); Rignot and Thomas (2002); Wu et al. (2002)
GRACE only	20	0.6	Wahr et al. (2000)
ICESat + GRACE	4–9	0.1–0.3	Wahr et al. (2000); Wu et al. (2002)
20year missions altimetry + gravity	3–4	0.1	Wahr et al. (2000)

At present, the data situation is changing significantly with the availability of precise temporal gravity variations from GRACE and altimetric ice surface data from ICESat and – in the next future – CryoSat. These missions allow to determine directly the ice volume and mass changes with a very high precision, with a nearly complete coverage of ice sheets and glaciers, with high spatial resolution and with sufficient temporal resolution (monthly sampling of mass changes by GRACE, whereas the time resolution of ice altimetry depends on the type of cross-over analysis). The lower rows of Table I show the improvements in accuracy expected from the use of these new data.

To derive the ice mass balance from these new data, one will have to deal with the following main problems:

- The most important task is the separation of signals from GIA, from recent ice mass changes and from the mass changes in the neighboring oceans, in atmosphere and hydrology.
- When volume changes from altimetry are converted to mass changes, ice compaction has to be modelled, which introduces considerable conversion errors. In particular, time variations of compaction caused by variable accumulation rates are a critical issue (Wahr et al., 2000; Wu et al., 2002).
- In the polar areas, sequences of years with higher-than-normal or lower-than-normal snowfall rates cause strong interannual variations in the ice mass balance, which are not yet well understood. Figure 2 (from Wahr et al., 2000) shows snow-ice variations from a climate model for a time span of 200 years, demonstrating that strong, fast changes alternate with much more stable periods. Since the current missions have a lifetime between 3 and 6 years, it becomes clear that they will capture rather a snapshot of current change, while for a reliable determination of interannual to secular variations longer mission durations have to be considered (Wahr et al., 2000). Wu et al. (2002) show that also for attempts to model the past ice load history by inversions, these interannual variations are the major error source.
- The capability of GRACE to detect variations in the ice mass balance is limited to a spatial resolution of about 500 km (spherical harmonic degree 40, see Figure 3). Variations over shorter scales will not be resolved.

Simulations by Wahr et al. (2000) and Velicogna and Wahr (2002a, 2002b) show, that a joint modelling and a separation of GIA signal and ice mass trend is possible. When using simulated GRACE data alone the separation does not succeed very well (Table I, second row), the achieved accuracies for ice mass and sea level change being not much better than from current data. When using GRACE and ice altimetry together, the accuracy improves considerably (Table I, third row), and some further improvements can be achieved adding a set of GPS vertical movement data. Also regional

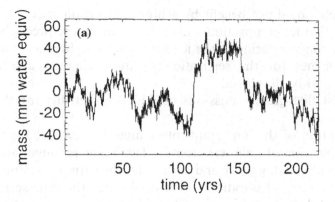

Figure 2. 219 years of monthly values of the snow-ice mass averaged over the Antarctic ice sheet, as predicted by the CSM-1 climate model, from Wahr et al. (2002). Units are equivalent water thickness.

variations could be recovered by these simulations, with accuracies varying between 5 and 20 mm/year of equivalent water thickness (Velicogna and Wahr, 2002a). In the approach of Wahr et al. (2000) and Velicogna and Wahr (2002a, 2002b), volume changes from ice altimetry are converted to mass changes, introducing a significant error due to the unknown ice compaction, which dominates the measurement errors. Therefore the results of

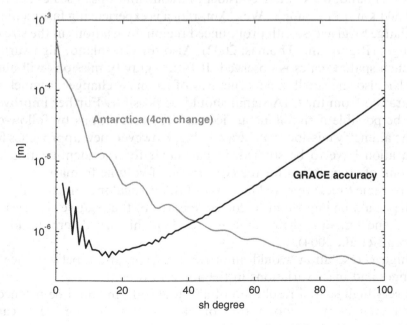

Figure 3. Degree amplitudes of the geoid effect of a 4 cm ice thickness change for the entire antarctic continent, compared to the GRACE baseline error.

this approach would not benefit by a higher precision gravity mission. For the future, either ice compaction must be modelled more precisely (e.g. based on extensive shallow coring of the ice sheets, see Huybrechts et al., 2004), or other approaches for the separation of mass changes and compaction changes have to be developed.

Future satellite gravity missions could bring benefits for the following areas:

- Determination of the important interannual ice mass changes (Figure 2) and identification of long-term trends. This requires longer and consistent time series of geoid, gravity and surface height variations. Future missions
- possibly with longer duration – would continue the time series started at present by GRACE and ice altimetry. Ideally, this continuation should be without interruption.
- Understanding of the Glacial Isostatic Adjustment mechanism and the viscoelastic properties of the Earth from gravity missions with better spatial resolution, higher accuracy and longer lifetime (Vermeersen, 2004). This will facilitate the separation of ice mass changes and mass shifts in the Earth's interior (Velicogna and Wahr, 2003a). Therefore, the mission precision requirements imposed by GIA research are supported by the needs of ice mass balance determination. GIA mass changes and ice mass changes have distinct spatial signatures. The GIA mass signal has most of its power at scales (wavelengths) greater than 500 km. The ice mass change signal, at the other hand, will contain considerable contributions at scales between 50 and 500 km. For example, West Antarctica is experiencing fast melting and discharge by glaciers, with pronounced regional variations in the size of the changes (Rignot and Thomas, 2002). Also for Greenland, high variability on such spatial scales is observed. If future gravity missions will allow to resolve also the small scale structures of ice mass changes, a much better separation from the GIA signal should be possible. Further improvement will be possible if simultaneous ice surface observations by follow-on satellite altimetry missions will be available. However, new approaches for the separation have to be studied, in particular for the identification of the amount of compaction for the conversion of volume to mass.
- At the same time, improved models of GIA vertical movements all over the globe would be important to correct records of the relative sea level at the coast and enhance their value for ice load history determination (Huybrechts et al., 2004).
- A higher resolution would improve the separation between ice mass changes and mass variations in the adjacent ocean.
- If a very high spatial resolution of about 20–30 km could be reached, this could even allow a monitoring of narrow fast flowing ice streams in Antarctica, which suddenly and unexpectedly stop or resume their movement (Joughin and Tulaczyk, 2002).

Therefore, a future mission should meet the following requirements:
- the time span covered by a single mission or series of missions should be as long as possible, at least 10 years;
- the mission accuracy should meet the needs of GIA research, i.e. in the order of 0.01 to 0.001 mm/year; (Vermeersen, 2004);
- the spatial resolution should reach 100 km or better.

Besides gravity time variations, also an improved static gravity field could contribute to study ice mass balance. The uncertainty of ice compaction mentioned above could possibly be reduced by combining gravity field and ice thickness data, see next section. Furtheron, a high resolution static geoid with a 10 cm accuracy or better could play a role for the determination of ice thickness along the grounding lines of ice shelves and glaciers that cover large parts of the coast of Antarctica. The geoid could be used as an approximation of sea level and allow the height determination of the ice bottom at the grounding line. From ice thickness, together with ice velocity determined by INSAR, the ice mass discharge can be assessed, which is an important contribution to ice mass balance and represents a major part to the present uncertainty in the mass budget of Antarctica (Huybrechts et al., 2004).

3. Ice Bottom Topography

The bottom topography beneath the ice sheets is an important boundary condition for ice flow models. Today it is obtained mainly from ice penetrating radar measurements. Current ice flow models use grids with 20 to 40 km resolution (Huybrechts et al., 2003). If the static gravity field from a future mission would reach such a resolution, it could allow an independent check of radar derived ice thickness. For such small spatial scales, airborne gravimetry is an important complementary data source. The better the spatial gravity field resolution achieved by a satellite mission, the broader will be the range of scales where both satellite and airborne data have good quality. This will enable a mutual validation of both sensor systems and a combination of results. Furthermore, when combining the geometry from radar with the ice mass derived from the gravity signal, ice density and compaction could be determined, which is today known only at few borehole sites. With simple rules of thumb one can derive that for an accuracy of 1 m in bottom topography at least a 0.1 mGal gravity accuracy is required. To detect a relative density anomaly of 0.001 for a 2 km thick ice sheet a 0.04 mGal gravity signal is required.

4. Sea Ice

The extent of the areas covered by sea ice and the transport of ice to lower latitudes by ocean currents play an important role for Earth climate and are an indicator of secular and interannual climate changes. Sea ice carries a big amount of freshwater and supplies cold water to the circulation system. These are important boundary conditions for ocean circulation. Another climate related issue is the increased reflection of solar radiation by sea ice. The distribution and thickness of sea ice is characterized by strong interannual variations, with important consequences for the high latitude ocean circulation and climate, which is shown by Venegas et al. (2001) for the Southern Ocean and by Laxon et al. (2003) for the Arctic.

While the extent of sea ice is being measured already today by various remote sensing techniques, the actual height of the ice surface and the freeboard height (the height of the sea ice at its edges) will be measured by the ICESat and CryoSat missions during the next years. To determine sea ice thickness and mass all over the sea ice cover, in addition to the ice surface also the geometry of the sea surface (as if there were no ice) has to be introduced. Therefore, the static geoid and the dynamic sea surface topography due to currents must be known. As the variations of the geoid variations are much larger than those of the sea surface topography, a static geoid in very high resolution down to wavelengths of about 10 km with an accuracy of 10 cm would lead to an improvement of an order of magnitude for sea ice thickness and mass transport determination (see Hvidegaard and Forsberg, 2002).

Acknowledgements

This work was funded in part by Deutsches Zentrum für Luft- und Raumfahrt (DLR) which is gratefully acknowledged.

References

Committee on Earth Gravity from Space: 1997, *Satellite Gravity and the Geosphere*, National Academy Press, Washington, D.C.

Huybrechts, P., Dietrich, R. Miller, H. and Haas C.: 2004, Ice mass balance and sea level, in Ilk, K.H., J. Flury, R. Rummel, P. Schwintzer, W. Bosch, C. Haas, J. Schröter, D. Stammer, W. Zahel, H. Miller, R. Dietrich, P. Huybrechts, H. Schmeling, D. Wolf, J. Riegger and A. Bardossy, A. Güntner (eds), *Mass Transports and Mass Distribution in the System Earth. Contribution of the New Generation of Satellite Gravity and Altimetry Missions to Geosciences.* TU München, GFZ Potsdam, pp.48–60.

Huybrechts, P.: 2002, *Quat. Sci. Rev.* **21**(1–3), 203–231

Hvidegaard S. M. and Forsberg R.: 2002, *Geophys. Res. Lett.* **29**(20), 1952, doi:10.1029/ 2001GL014474.

James, T. S. and Ivins E. R.: 1997, *J. Geoph. Res. Vol.* **102**(B1), 605–634, doi:10.1029/ 96JB02855.

Joughin, I. and Tulaczyk S.: 2002, *Science*, **295**, 476–480.

Laxon, S., Peacock, N. and Smith D.: 2003, *Nature*, **425**, 947–950.

Rignot, E. and Thomas R.H.: 2002, *Science*, **297**, 1502

Thomas, R. H.: 2001, *EOS Transactions*, AGU **82**(34), 369–373.

Velicogna, I. and Wahr J.: 2002a, *J. Geophys. Res.* **107**(B10), 2263, doi:10.1029/2001JB000708.

Velicogna, I. and Wahr J.: 2002b, *J. Geophys. Res.*, **107**(B10), 2376, doi:10.1029/2001JB001735.

Venegas, S., Drinkwater, M.R. and Schaffer G.: 2001, *Geophys. Res. Lett.* **28**(17),3301–3304.

Vermeersen, L. L. A.: 2004, *Challenges from Solid Earth Dynamics for Satellite Gravity Field Missions in the Post-GOCE Era*. This issue.

Wahr, J., Wingham, D. and Bentley C.: 2000, *J. Geophys. Res.* **105**(B7), 16279–16294

Wahr, J. and Velicogna I.: 2003, *Space Sci. Rev.* **108**(1), 319–330.

Woodworth, Ph.: 2004, *Global Sea Level Change*. This issue.

Wu, X., Watkins M.M., Ivins E.R., Kwok R., Wang, P. and Wahr J.: 2002, *J. Geophys. Res.* **107**(B11), 2291, doi:10.1029/2001JB000543.

Earth, Moon, and Planets (2005) 94: 93–102
DOI 10.1007/s11038-004-5172-9

BENEFITS TO STUDIES OF GLOBAL SEA LEVEL CHANGES FROM FUTURE SPACE GRAVITY MISSIONS

PHILIP L. WOODWORTH

Proudman Oceanographic Laboratory, Joseph Proudman Building, 6 Brownlow Street, Liverpool L3 5DA, UK (E-mail: plw@pol.ac.uk)

(Received 10 August 2004; Accepted 21 October 2004)

Abstract. Global sea level rise will present a major scientific, environmental and socio-economic challenge during the 21st century. This paper reviews the main oceanographic and geophysical processes which contribute to sea level change, with particular emphasis on the ability of space gravity missions to contribute to an enhancement of our understanding of the various processes, and ultimately to a better understanding of sea level change itself. Of special importance is the need to understand better the ocean circulation, and the contribution of ocean thermal expansion to sea level change.

Keywords: Climate change, gravity field, ocean circulation, sea level changes

1. Introduction

The study of long-term changes in sea level is of great scientific interest and considerable practical importance to the environmental and economic infrastructure of coastal zones. The recent reports of Church et al. (2001) and Woodworth et al. (2004) have provided overviews of the scientific issues connected to, and the coastal impacts of, the sea level changes of the past century and of the next 100 years.

Previous working groups have demonstrated the potential for space gravity missions to provide information on the spatial and temporal dependence of the Earth's gravity field, which can lead to improvements in the scientific understanding of a number of processes contributing to sea level changes. In particular, Balmino et al. (1999) presented the benefits of the considerably improved knowledge of the geoid from the GOCE spatial gravity mission to studies of the ocean circulation, solid Earth, glaciological processes, geodesy and satellite orbit determination, which together should lead an improved understanding of sea level change. The case for a temporal gravity mission such as GRACE was also constructed partly around the topic of sea level change, by providing better understanding of the global hydrological cycle and of the ocean thermohaline circulation together with processes in the solid Earth, notably Glacial Isostatic Adjustment (GIA, see NRC, 1997; GRACE, 1998). The spatial and temporal accuracies which these missions are expected to achieve in order to meet their scientific objectives are summarised in Rummel (2003).

Although the cases for these two missions were based to a great extent around the need for greater understanding of sea level change, it is important to realise that, just as sea level change occurs as a result of many changes in the earth's climate and geosphere (e.g. see Figure 1), so the benefits of space gravity to sea level science will be realised only after each of the above-mentioned science issues has benefited individually. Therefore, it is difficult to state quantitatively at the moment just how great the benefits to sea level studies will be.

Nevertheless, we can be confident about one aspect of future sea level change in that it will contain a significant contribution from climate change ("global warming") resulting in the thermal expansion of the ocean. Processes such as thermal expansion are modelled within Atmosphere Ocean General Circulation Models (AOGCMs). From the Intergovernmental Panel on Climate Change Third Assessment Report (IPCC TAR) (Church et al., 2001), we learn that AOGCMs predict a sea level rise of between 9 and 88 cm between 1990 and 2100, with a central value of 48 cm. This wide range is obtained because a full set of emission scenarios were used by the TAR, together with a number of AOGCM formulations. The mid-range estimate represents a rate of rise in the 21st century of approximately 2–4 times that of the past 100 years.

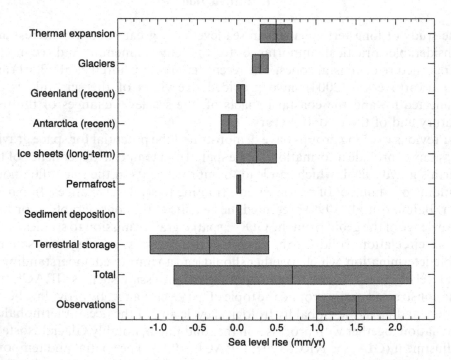

Figure 1. Ranges of uncertainty for the average rate of sea level rise during 1910–1990 and the estimated contributions from different processes (Church et al., 2001).

Figure 2 is an updated and corrected version of Figure 11.11 of Church et al. (2001). It demonstrates the wide range of predictions from AOGCMs (see also other figures and tables in Church et al., 2001 and Woodworth and Gregory 2003). However, even though their predictions differ greatly, the majority of the future rise in each is due to thermal expansion. For example, from the HadCM3 AOGCM one predicts a thermal ezxpansion for 1990–2100 of 24 cm for the IS92a scenario, which will occur within an overall rise of between 18 and 46 cm if all contributing terms are considered.

Consequently, the first difficult but most important question is, how will the sea level predictive capability of Atmosphere Ocean General Circulation Models (GCMs) improve, and the range of uncertainty narrow, as a consequence of the gravity missions, and, in particular, how will their predictions of thermal expansion improve? This issue was addressed most recently by Woodworth and Gregory (2003).

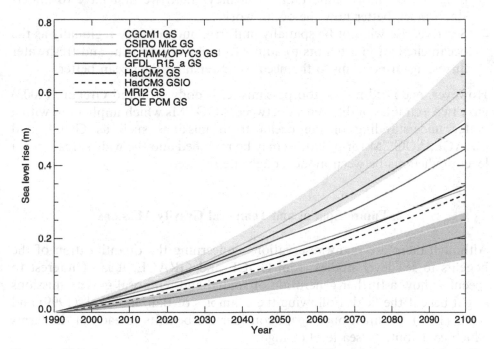

Figure 2. Global average sea level rise 1990–2100 for the IS92a emission scenario, including the direct effect of sulphate aerosols. Thermal expansion and land ice changes were calculated from AOGCM experiments, and contributions from changes in permafrost, the effect of sediment deposition and the long-term adjustment of the ice sheets to past climate change were added. For the models that project the largest (CGCM1) and the smallest (MRI2) sea level change, the shaded region shows the bounds of uncertainty associated with land ice changes, permafrost changes and sediment deposition. Uncertainties are not shown for the other models, but can be found in Table 11.14 of Church et al. (2001). The outermost limits of the shaded regions indicate the range of uncertainty in projecting sea level changes for the IS92a scenario.

Another conclusion from the IPCC TAR modelling is that a future sea level rise from either all the contributing terms, or from thermal expansion alone, will not be spatially uniform, but will vary regionally as the ocean circulation attempts to adjust to the changing fluxes. This may have important consequences for particular regions, if they experience significantly greater rises than the global-mean. Unfortunately, while the various AOGCMs agree that major spatial variations will occur, they disagree on the exact geographical pattern.

In summary, two conclusions can be drawn from Church et al. (2001) and Woodworth and Gregory (2003):

- Global-mean thermal expansion needs to be very well understood for predicting the sea level rise of the next 100 years, and in particular for the second half of the 21st century as the effects of climate change become progressively more important. To achieve that, we first have to understand much better how the ocean works.
- Sea level rise will not be spatially uniform, but will vary regionally as the ocean circulation attempts to adjust to the changing heat and freshwater fluxes, again pointing to the need to understand the ocean better.

However, one need not be too pessimistic. Woodworth and Gregory (2003) gave two examples of differences between AOGCMs which imply that, with a better understanding of the ocean from missions such as GOCE and GRACE, AOGCM formulations may be modified and the wide spread in sea level predictions between models might be reduced.

2. Future Spatial and Temporal Gravity Missions

Although there are many reservations concerning the quantification of the benefits to sea level studies from GOCE and GRACE, it is of interest to speculate how a further generation of spatial and temporal gravity missions might benefit the field. Following the examples of Balmino et al. (1999) and NRC (1997), we must once again consider the benefits to each of the terms which contribute to sea level change.

2.1. BENEFITS FROM IMPROVED RESOLUTION OF A SPATIAL GRAVITY MISSION

The steady-state ocean circulation
The case for GOCE was constructed around its ability to provide a measurement of the geoid with centimetric accuracy down to spatial scales of 100 km half-wavelength, which corresponds to a typical deep ocean Rossby

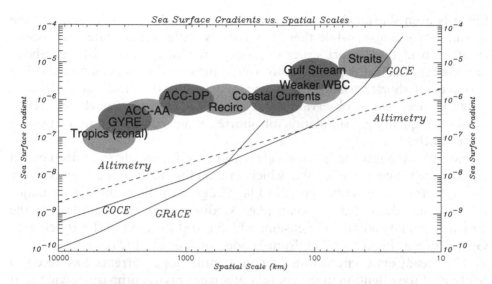

Figure 3. Highly schematic illustration of sea surface gradients (relative to the geoid) of several components of the ocean topography compared to mean sea surface slope accuracy from altimetry (dashed line) and to geoid slope accuracy from space gravity missions such as GOCE (thick line) and GRACE (thin line). "Gulf Stream" represents the stronger deep ocean fronts including those of the Gulf Stream itself and of, for example, the Antarctic Circumpolar Current. "Recirc" represents the Gulf Stream recirculation. "Weaker WBC" represents the weaker Western Boundary Currents (e.g Brazil Current) with spatial scales of order 100 km and gradients of order few 10^{-6}. "ACC-DP" and "ACC-AA" represent a major current such as the ACC at Drake Passage or at the wider African and Australian choke points respectively. "GYRE" represents a typical 1 m ocean gyre over 3000 km scale. "Coastal currents" represents the myriad of coastal currents, flows through longer straits and meridional equatorial signals with space scales of order 100 km and gradients of 10^{-6}. "Straits" represents flows through short straits which are at the limit of spatial resolution. Note that at very long wavelengths, where GRACE accuracy is superior to that of GOCE, remaining altimeter orbit and other systematic uncertainties are still significant. (from Woodworth et al., 1998).

radius of deformation. The case was made that spatial scales of the steady-state ocean surface circulation larger than the radius can be considered to be in approximate geostrophic balance with the sea surface topography measured by means of an altimetric mean sea surface from which a precise geoid can be subtracted. Therefore, the steady-state circulation can be determined from the two data sets relatively straightforwardly. The 100 km scale was also chosen on the basis that in present deep ocean models one observes relative little steady state circulation at spatial scales shorter than the 100 km (cf. Figure 3.11f of Balmino et al., 1999; see also Woodworth et al., 1998; Le Provost and Bremond, 2003).

GOCE should indeed provide a measurement of the geoid which will be good enough for deep ocean studies down to that scale. However, one might now revisit the benefits to knowledge of the circulation through improvements towards even shorter deep ocean scales and towards shallower areas of the ocean. For example, while there is relatively little steady state deep ocean circulation at these short scales in models, it clearly does exist (see above references) even if it is probably not well represented in present models. The existence of shorter scale features (meanders, fronts) is also known from a range of deep ocean observations. As higher resolution models are developed, the quality of simulation of shorter scale features should improve significantly.

Notable amongst the shorter scale features of the steady state deep ocean circulation which do exist and which are undoubtedly important is flows through straits (e.g. Gibraltar, Florida, Skagerrak). Such flows play major roles in the adjacent deep ocean (e.g. Mediterranean deep outflow in the North Atlantic) and have dimensions which are at the limit of the anticipated GOCE range (Figure 3 taken from Woodworth et al., 1998).

The steady-state circulation of continental slope currents has received relatively little attention in the context of space gravity, primarily owing to it also being at the spatial limit of GOCE, with currents typically several 10s-km wide. However, most continental slopes have currents with transports of several Sv ($1 \ Sv = 10^6 \ m^3 \ s^{-1}$). This is small in comparison to the approximately 100 Sv transports of the Gulf Stream and the other deep ocean systems. However, they together comprise a major component of the global ocean circulation and are of local importance with regard to shelf-deep water fluxes. An example of the possible improvement in oceanographic knowledge of the NW European slope current from satellite altimetry together with state-of-the-art geoid models was provided by Haines et al. (2003). Flows on the continental shelves can also be of interest. For example, in the North Sea the effective Rossby radius can be very short (several to 10s km) with flows from rivers (notably the Rhine) onto the shelf being essentially steady state at this spatial scale and with a sea surface topography signature of order 10 cm.

While these short-wavelength signals could be significant within discussions of the overall benefits of a new spatial gravity mission, especially with regard to regional ocean modelling, it is less clear how importantly this aspect of the steady state circulation might benefit sea level studies.

Ice sheets (Greenland and Antarctica). Balmino et al. (1999) referred to the limited benefits to knowledge of ice sheet dynamics, and consequently of mass balance, from GOCE, given that a < 50 km resolution bed geometry was needed via measurement of the geoid at that scale for study of Antarctic ice streams. In addition, there was little benefit to knowledge of the bed geometry of Greenland which is well-known from radar sounding. Therefore, a new spatial gravity mission would have to provide a resolution of order 20 km to be of use.

Glacial isostatic adjustment. Balmino et al. (1999) implied that the spatial gravity provided by GOCE would be as specified by the science requirements for anticipated GIA studies. Therefore, there is as yet no case for a further, higher resolution spatial gravity mission for this topic.

Tectonics. Balmino et al. (1999) demonstrated that understanding could be increased of mechanisms of active tectonics in areas such as Italy and the Adriatic, with data from GOCE enabling the accurate determination of rates of vertical land movement from tectonics, to which rates from GIA can be added, for comparison to rates observed from GPS or tide gauge data (Di Donato et al., 1999; Woodworth, 2003). The modelling of such tectonic processes would clearly benefit from higher resolution spatial gravity information.

Hydrology. The uncertainty in the net contribution of hydrological processes to global sea level change during the 20th century is very large and is a major point of discussion when considering the combination of various terms which make up the observed 20th century sea level rise (Figure 1). However, a spatial gravity mission such as GOCE or its successor is not suited to address this issue.

Glaciers. Similarly, Balmino et al. (1999) did not consider that GOCE could contribute to studies of mass balance of mountain glaciers, the understanding of which it was considered would be undertaken by various forms of altimetry and other types of remote sensing in future.

Bathymetry. Balmino et al. (1999) noted that, at the 10-500 km spatial scale, spatial gravity can be directly related to ocean bathymetry. A lack of knowledge of the bathymetry of the global ocean is a major factor in the construction of ocean numerical models and bathymetric information recovered to 10 km resolution from spatial gravity would be very desirable.

Geodetic issues. There are a number of sea level-related geodetic issues such as the need for a geoid to use within "GPS-levelling" between tide gauges, and the need to remove gravity field errors in the orbits of altimeter satellites. The former was discussed by Balmino et al. (1999). The need for local *in situ* gravity in an area approximately 50 km around tide gauges, in addition to data from GOCE, is noted if geoid omission errors are to be removed to the cm level, as required for GPS-levelling between gauges. Balmino et al. (1999) considered that the gravity field errors in altimeter orbits would be essentially zero after GRACE and GOCE.

2.2. BENEFITS FROM IMPROVED SPATIAL/TEMPORAL RESOLUTION OF A TEMPORAL GRAVITY MISSION

Deep ocean circulation. NRC (1997) and GRACE (1998) have demonstrated the potential for GRACE to provide monthly maps of gravity changes with an accuracy corresponding roughly to a millimetric disk of water of spatial scale typically 700 km. A temporal gravity mission such as this will be of major benefit to ocean circulation studies by providing a monitor of mass transports, complementing the monitoring of sea surface height by altimetry. Recent papers have also demonstrated the importance of understanding ocean mass transfers to studies in geophysics, such as in accounting for recent variations in J2 (Dickey et al., 2002).

Discussion of the utility of temporal gravity to oceanography beyond GRACE is provided elsewhere in this volume. The insight into ocean circulation gained will eventually be included in AOGCMs to the particular benefit of the thermal expansion component of sea level studies as discussed above.

Coastal ocean circulation. Benefits in this area are not so clear. The effective spatial resolution would need to improve over GRACE by an order of magnitude, while preserving a similar temporal resolution. While further studies could be undertaken from a purely coastal oceanography perspective, it is unlikely that benefits would extend directly to sea level studies.

Glacical isostatic adjustment. Wahr and Velicogna (2003) have discussed the decoupling of GIA from other processes within GRACE data (e.g. decoupling the Greenland GIA signal from the present-day elastic term, and from neighbouring ocean mass changes). The authors conclude that effective decoupling will result in improved estimates of the Earth's viscosity profile, although errors in ice history models remain a major issue. Higher resolution would benefit the decoupling.

Hydrology. As discussed above, hydrological changes was a major issue in discussion of reasons for 20th century sea level change (Church et al., 2001; Woodworth and Gregory, 2003; Cazenave et al., 2003). During the 2002 Bern Symposium on Space Gravity, Wahr discussed the spatial dealiasing of GRACE data in order to provide time series of groundwater mass for (large) drainage basins. Higher resolution would benefit the construction of such hydrological time series.

Ice sheets (Antarctica and Greenland) and Glacier Groups (e.g. Alaska). Ice sheet mass changes should be observable from a mission such as GRACE to the same spatial/temporal resolution as for ocean or atmospheric mass and hydrology. Interpretation in terms of sea level change will take place in combination with altimetric measurements. Low latitude glaciers contain ice equivalent to approximately 40 cm of sea level. Alaska's glaciers, which can be considered as comprising a small ice sheet, are known to be melting fast and contributing to current sea level change (Arendt et al., 2002; Dickey et al., 2002). Once again, the effective monitoring of such mass changes, in combination with altimetry, would benefit sea level studies.

In addition to considering the sea level-equivalent of melt water from ice sheets and glaciers, the signal observed in tide gauge records from the melting will include a solid earth loading term (Tamisiea et al., 2001). To model loading well, the spatial distribution of mass change needs to be known as well as possible. Consequently, the 'fingerprint methods' such as those of Tamisiea et al. will benefit from higher temporal gravity resolution.

Global-mean mass monitoring. Nerem and Leuliette (2003) presented simulations showing that GRACE should be able to monitor non-secular, monthly average changes in ocean mass of approximately 3 mm water-equivalent with a spatial resolution of order 500 km, comparable to previous estimates. They also suggested that GRACE would be able to monitor changes in global-mean ocean mass to approximately 1 mm.

These simulations are very encouraging, but the authors stress that secular signals from GIA and melting polar ice result in major difficulties with regard to determination of secular ocean mass change, and such ambiguity would be reduced with a higher resolution mission. Multiple satellites, and combined analyses with altimetry, could reduce such aliasing in time series. The capability of providing millimetric accuracy time series of global-mean ocean mass (an order of magnitude more precise than current time series of global-mean sea level) would clearly be an important addition to sea level studies.

3. Conclusions

It can be seen that both a higher resolution spatial gravity mission and a higher resolution temporal gravity mission would provide benefits to sea level studies beyond those anticipated from GOCE and GRACE. However, in spite of the undoubted benefits of an improved gravity field from a spatial gravity mission (to, say, 1 cm accuracy over 20 km half-wavelength rather than 100 km from GOCE), an improved temporal gravity mission would appear to have higher priority. As explained by Rummel (2003), we need to understand mass transports between atmosphere, oceans, hydrology, glaciology etc., with the study of sea level change forming part of that wider requirement, and that can be addressed best in the medium term by further temporal gravity missions, with an enhanced spatial gravity mission later (the GOCE studies resulted in a recommendation for missions of the GOCE type to be repeated at perhaps decadal intervals).

Finally, we can return to the major question posed by Woodworth and Gregory (2003), and ask how the predictability of sea level changes within AOGCMs will benefit from the new missions. This is still a difficult question to answer quantitatively. However, if we knew global fluxes significantly better than we know them now, then we would be able to parameterise them better in the coarse resolution AOGCMs, and thermal expansion in particular should be handled much better.

References

Arendt, A.A., Echelmeyer, K.A., Harrison, W.D., Lingle, C.S. and Valentine, V.B.: 2002. Rapid wastage of Alaska glaciers and their contribution to rising sea level. Science, **297**, 382–386.

Balmino, G., Rummel, R., Visser, P. and Woodworth, P.: 1999. Gravity Field and Steady-State Ocean Circulation Mission. Reports for assessment: the four candidate Earth Explorer Core Missions. European Space Agency Report SP-1233(1), 217pp.

Cazenave, A., Cabanes, C., Dominh, K., Gennero, M.C. and Le Provost, C.: 2003. Present-day sea level change: observations and causes. Space Sci. Rev. **108**, 131–144.

Church, J.A., Gregory, J.M., Huybrechts, P., Kuhn, M., Lambeck, K., Nhuan, M.T., Qin, D. and Woodworth, P.L.: 2001, Changes in sea level. In: J. T. Houghton, Y. Ding, D. J. Griggs, M. Noguer, P. J. van der Linden, X. Dai, K. Maskell and C. A. Johnson (eds.), Climate Change 2001: The Scientific Basis. Contribution of Working Group I to the Third Assessment Report of the Intergovernmental Panel on Climate Change. Cambridge: Cambridge University Press. 881pp.

Dickey, J.O. Marcus, S.L., de Viron, O., and Fukumori, I.: 2002. Recent Earth oblateness variations: unraveling climate and postglacial rebound effects. Science, **298**, 1975–1977.

Di Donato, G., Negredo, A. M., Sabadini, R., and Vermeersen, L. L. A.: 1999. Multiple processes causing sea-level rise in the central Mediterranean. Geophysical Research Letters, **26**, 1769–1772.

GRACE, 1998. Gravity Recovery and Climate Experiment science and mission requirements document. Revision A, JPLD-15928, NASA's Earth System Science Pathfinder Program.

Haines, K., Hipkin, R., Beggan, C., Bingley, R., Hernandez, F., Holt, J., Baker, T. and Bingham, R.J. 2003. Use of altimetry and in situ gravity data for coastal dynamics studies. Space Sci. Rev. **108**, 205–216.

Le Provost, C. and Bremond, M. 2003. Resolution needed for an adequate determination of the mean ocean circulation from altimetry and an improved geoid. Space Sci. Rev. **108**, 163–178.

Nerem, R. S. and Leuliette, E. W. 2003. Measuring the distribution of ocean mass using GRACE. Space Sci. Rev. **108**, 331–334.

NRC, 1997. *Satellite Gravity and Geosphere.* National Academy Press, Washington, D.C.

Rummel, R.: 2003. How to climb the gravity wall. Space Sci. Rev. **108**, 1–14.

Tamisiea, M. E., Mitrovica, J. X., Milne, G. A., and Davis, J. L.: 2001. Global geoid and sea level changes due to present-day ice mass fluctuations. J. Geophysical Res. **106**, 30849–30863.

Wahr, J. and Velicogna, I.: 2003. What might GRACE contribute to studies of post glacial rebound? Space Sci. Rev. **108**, 319–330.

Woodworth, P. L., Johannessen, J., Le Grand, P., Le Provost, C., Balmino, G., Rummel, R., Sabadini, R., Suenkel, H., Tscherning, C. C., and Visser, P.: 1998. Towards the definitive space gravity mission. Int. WOCE Newslett. **33**, 37–40 and 24.

Woodworth, P. L., Gregory, J. M., and Nicholls, R. J.: 2004, Long term sea level changes and their impacts, in A. Robinson and K. Brink (eds.), The Sea, **13**, 717–752, Harvard University Press.

Woodworth, P. L., and Gregory, J. M.: 2003, Benefits of GRACE and GOCE to sea level studies. Space Sci. Rev. **108**, 307–317.

Woodworth, P. L.: 2003, Some comments on the long sea level records from the northern Mediterranean. J. Coast. Res. **19**, 212–217.

Earth, Moon, and Planets (2005) 94: 103–111
DOI 10.1007/s11038-005-3245-z

GRAVITY AND TOPOGRAPHY OF MOON AND PLANETS

R. RUMMEL

Institut für Astronomische und Physikalische Geodäsie, TU München, Germany
(E-mail rummel@bv.tum.de)

(Received 6 October 2004; Accepted 4 March 2005)

Abstract. Planetology serves the understanding on the one hand of the solar system and on the other hand, for investigating similarities and differences, of our own planet. While observational evidence about the outer planets is very limited, substantial datasets exist for the terrestrial planets. Radar and optical images and detailed models of gravity and topography give an impressive insight into the history, composition and dynamics of moon and planets. However, there exists still significant lack of data. It is therefore recommended to equip all future satellite missions to the moon and to planets with full tensor gravity gradiometers and radar altimeters.

Keywords: Gradiometry, gravity, moon, planets

1. Moon and Planets

Planetology serves two main purposes. First it is aimed at a deeper understanding of our solar system, its history, its characteristic features and its expected future development. Secondly, from comparison and from investigating similarities and differences, the study of planets helps in understanding our own planet. The major planets of our solar system are usually divided into two types. The outer planets, Jupiter, Saturn, Uranus and Neptune contain most of the mass of the planetary system. They have a large distance from the Sun, large diameters, low density, low surface temperature and well-developed satellite systems. The inner planets, Mercury, Venus, Earth and Mars, are referred to as terrestrial planets. They are much closer to the Sun, have small diameters, high density and less developed satellite systems or no moons at all. Compare Watts (2001).

The terrestrial planets are similar in many respects, yet there are significant differences among them. All four seem to be formed from mass accretion in the solar nebula. Active plate tectonics and the large height difference between the old continental crust and the young oceanic crust, however, are assumed to be unique features of the Earth. Mercury, and our Moon, have a continuous lithosphere, with their surface largely characterized by volcanism and impact craters. Typical for Mars is its hemispheric dichotomy. Its Southern hemisphere is densely cratered whereas the surface of the Northern hemisphere consists of lightly curved plains. Its surface is modified in its early history by atmospheric influences and flow of presumably water. Venus, the planet most similar to the Earth, is cloud covered; its surface is relatively hot. Also Venus is characterized by volcanism. See Schubert et al. (2001).

The primary sources of information about the planets are imaging of the surface (and telescope observations in the past) and orbit analysis from planetary fly-bys, as in the case of the outer planets and of a number of planetary moons or asteroids, or from elliptical orbits of planetary orbiters. For some planets, detailed topographic information exists, too, based on photogrammetric or altimetric measurements. Only in the case of Mars and Moon, rock samples could be analyzed in-situ or after being returned to the Earth, respectively. Gravity field information from fly-bys provides some elementary information such as mass, axiality and angularity; compare Kaula (1991). Already the combined use of images, displaying all characteristic surface features, and of gravity gives a wealth of information about the roughness or smoothness of the field and about the correlation of gravity variations with the topographic relief. The addition of topographic models is essential, however, for a deeper understanding of the evolution, thermal history, tectonics, the state of isostatic compensation and many more fundamental aspects of planetology. Planetology based on gravity and topography is in many ways comparable to a situation in solid Earth geophysics at the beginning of the 20th century.

2. Moon

The two typical visual features of the Moon's surface are the lighter looking, elevated highlands and the much darker maria. The latter are large impact basins, most of them filled by basalts. Another characterization results from the systematic difference in elevation between the far-side and the near-side. The far-side has much less and less flooded maria. Furthermore far-side elevations are systematically higher with the exception of the South-Pole-Aitken basin. The latter is a large impact basin with little basaltic fill. It is the lowest-lying region on the Moon. Topographic heights on the Moon range from -8 km to $+8$ km relative to a mean elevation. The evolution of the Moon can be divided into three phases: highland formation before 4 Ga (formed in the early Moon history and crystallized from a global magma ocean), mare formation between 3.8 and 4.0 Ga (bombardment resulting in many, large and deep basins, later filled by basalts, similar in composition to Earth oceanic basalts) and surface quiescence, (Schubert et al., 2001). It is assumed that after impact, the basins were filled with basalts to a hydrostatic level, which is above the original basin floors on the near-side and below basin floors on the far-side. Heating and weakening of the crust resulted in mantle rebound, in uplift of the crust-mantle boundary and in the generation of "plugs", (Konopliv et al., 1998).

3. Gravity and Topographic Models of the Moon

Rather detailed spherical harmonic topographic and gravity models exist for the Moon. Gravity models exist up to degree and order 165. They are based on data from Lunar Orbiters 1 to 5, Apollo 15 and 16, Clementine and, in particular, Lunar Prospector. In its final phase, the orbital altitude of Lunar Prospector was as low as 25 km. Examples of lunar gravity models are:
– Goddard Lunar Gravity Models 1 and 2 (GLGM-1 and 2)
– Lunar Prospector Models (LP75n, LP100n, LP165P).

See Figure 1 for the RMS gravity power per degree of LP165P and Floberghagen (2002).

A weakness of all existing lunar gravity models is the lack of directly observed far-side data. By means of a sophisticated approach based on the analysis of accelerations of orbit gravitational perturbations accumulated at the end of its occultations, the effect of lack of far-side data is somewhat

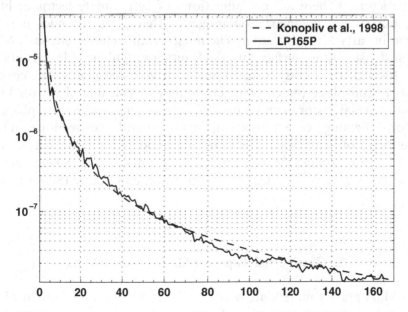

Figure 1. RMS gravity power per degree of lunar gravity model LP165P up to degree and order 165, as well as the power law by Konopliv et al. (1998).

diminished now-a-days, (Nerem, 1995; Konopliv et al., 1998). The empirical power law derived for the Moon gravity (Konopliv et al., ibid and Figure 1) is approximately

$$\sigma_n = 1.2 \cdot 10^{-4}/n^{1.8},$$

where σ_n approximates the RMS per degree

$$\sigma_n = \sqrt{\frac{1}{2n+1}\sum_{m=0}^{n}(\bar{C}_{nm}^2 + \bar{S}_{nm}^2)}.$$

It indicates that the Moon is closer to equilibrium than the Earth. Topographic models are based on Clementine lidar altimeter data and are developed up to degree and order 90. Based on the available gravity and topographic models, the degree of isostatic compensation can be inferred globally and locally for individual mares; compare Konopliv et al. (1998) and Watts (2001). Strong positive gravity anomalies are associated with mass concentrations in the near-side mare basins, the so-called mascons.

4. Mars

The surface of Mars is characterized by a wide variety of volcanic and tectonic structures. There are no indications of active plate tectonics like on Earth, but crustal magnetization indicates that plate tectonics may have occurred in early Mars evolution. The most striking characteristic of Mars is the clear division of its surface into the Southern highlands (densely cratered, rough, formed in its early history) and into volcanic plains that cover the Northern hemisphere (much younger, similar to volcanic plains on Venus). Martian thermal history can be divided into a very active early phase with accretional heating, core formation, strong mantle convection, and high surface fluxes of heat and magma and a second phase – the last 3.5 Gyr – marked by slow cooling. Mantle plumes play a major role in heat exchange. Very likely its core is completely fluid and non-convecting (Schubert et al., 2001).

5. Gravity and Topographic Models of Mars

Recent Mars gravity models are based on Doppler radio tracking of Mariner 9, Viking-1 and 2 and, primarily Mars Global Surveyor. Fields were developed by NASA GSFC and JPL and solved up to degree and order 80 or 85;

compare Smith et al. (1999) and Lemoine et al. (2001). Maximum values of free air anomalies are one order of magnitude higher (3000 mGal) than on Earth. The areoid exhibits a very distinct hemispheric East–West division with a range from −800 m to +2000 m, again more than an order higher in magnitude than on Earth. According to Smith et al. (1999), the approximate power law of Mars gravity field is, compare Figure 2:

$$13 \cdot 10^{-5}/n^2.$$

Topographic models are based on the Mars Orbiter Laser Altimeter (MOLA) with an elevation precision of about 13 m and an expansion up to degree and order 60, corresponding to 178 km. The topographic map shows the strong North–South dichotomy with a strong slope from South to North (Smith et al., 1999; Zuber et al., 2000).

Determination of Bouguer anomalies from gravity and topography provides a first look into subsurface Mars variations. Under the assumption of a single subsurface interface, a constant density contrast between crust and mantle and a radially uniform core and mantle density, Zuber et al. (2000) translated the Bouguer anomalies to crustal thickness. They found an average thickness of 50 km and variations in crustal thickness between 3 km and 92 km. A fundamental open issue is the formation of the northern lowlands. Based on the results of gravity and topography, they (ibid) consider an

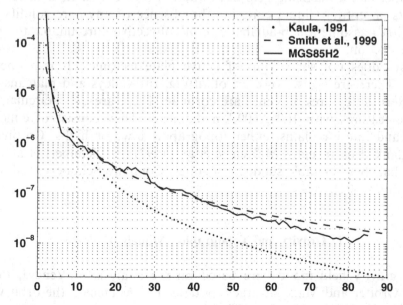

Figure 2. RMS gravity power per degree of the Mars Global Surveyor gravity model MGS85 1+2 up to degree and order 85 (Sjogren, 2002), as well as the power laws by Kaula (1991) and Smith et al. (1999).

impact hypothesis unlikely. Instead, either volcanic or sedimentary fill is favored. Furthermore Zuber et al. (2000) have come to the conclusion that negative gravity lineations as well as the geometric characteristics and the flow properties of these structures indicate either subaerial or subaqueous flow. Unlike the Earth, there is a high correlation between gravity and topography for Mars for the lower spherical harmonic degrees. Whereas ongoing geodynamic processes related to mantle convection, subduction and post-glacial rebound mainly determine the lower harmonics of the Earth's geoid, topographic signatures related to volcanic activity in the past like the Tharsis complex (Olympus Mons) seem to mainly determine the lower harmonics of Mars' areoid (Vermeersen, private communication).

6. Venus

Venus is the planet that most closely resembles Earth, in terms of size, mass and density. Yet there are substantial differences. Its surface topography is much smoother; Venus has a CO_2-rich atmosphere and surface temperatures of several hundred degrees. Gravity anomalies are smaller than those of Moon and Mars and they correlate better with topography than on Earth. This suggests at least partial isostatic compensation.

Venus has no intrinsic magnetic field. Schubert et al. (2001) suggest that the planet had a magnetic field until about 1.5 Gyr ago. Its absence today suggests an entirely liquid core, i.e. the absence of any core solidification. They (ibid) assume a sub-adiabatic, non convective core unable to sustain dynamo action.

Venus does not show signs of active plate tectonics. It is a one-plate planet. There are signs, however, of rifting, rift valleys and plateaus. Very characteristic features of Venus are coronae. These are quasi-circular topographic features with 100–2,600 km diameter. They consist of concentric ridges and interior plains, either topographic lows or highs. Coronae are often flanked by troughs. McKenzie et al. (1992) and Schubert et al. (1994) argue that they resemble subduction zones. Compare Schubert et al. (2001) and Watts (2001).

7. Gravity and Topographic Models of Venus

Venus gravity field models are based on the analysis of Venera, Pioneer Venus Orbiter and Magellan tracking data. For Magellan, the orbit was in 1993 circulized by aero-braking. This resulted in a significant improvement of gravity modeling (Nerem, 1995). Maximum resolution is degree and order

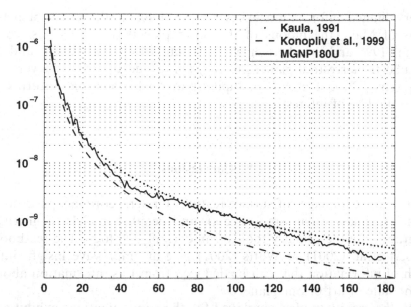

Figure 3. RMS gravity power per degree of the Venus gravity model MGNP180U (Sjogren et al., 1997), as well as the power law by Kaula (1991) and Konopliv et al. (1999).

180 (Konopliv et al., 1999), see Figure 3. The approximate power law of Venus gravity field is, due to Konopliv et al. (ibid),

$$1.8 \cdot 10^{-5}/n^{2.3}.$$

Very detailed topographic mapping has been obtained from Magellan, too; compare (Rappaport et al., 1999).

The key question addressed on basis of the available radar images, gravity and topographic models is the type of thermal evolution on Venus, (Kiefer and Potter, 2000; Schubert et al., 2001). Is there a vigorous mantle convection associated with a thin lithosphere and heat flow comparable to that on Earth, or is there a thick lithosphere with low heat flow? Schubert et al. (ibid) analyze the strong positive correlation of the "geoid" and topography at long wavelengths. It would result in compensation depths as deep as 200–300 km, while the actual lithospheric thickness is below 40 km. This suggests dynamic compensation by mantle convection. Kiefer and Potter (ibid) analyze gravity anomalies of several large shield volcanos. They use the superposition of several axis-symmetric Gaussian loads based on a least-squares fit to topographic data. The resulting elastic lithospheric thicknesses are between 10 and 22 km. These estimates are also strong functions of the lithospheric thermal gradient. Also their conclusions favor the hypothesis of a thin lithosphere for Venus.

Very limited information exists about Mercury, the outer planets, the moons of Mars, the moons of the outer planets and asteroids, compare (Nerem, 1995 and Schubert et al. 2001, ch. 14.1.1 and 15.9.4. and 5). In particular the terrestrial planet Mercury being a dead planet, the very active, though small Jupiter moon Io, Jupiter's moon Europa and Saturn's moon Titan would be of interest.

8. Conclusions

Radar or optical images of surface features (craters, coronae, plateaus, rifts, ridges, roughness versus smoothness), and detailed models of gravity and topography are the primary tools of modern planetology. The textbooks by Schubert et al. (2001) or Watts (2001) give an impressive insight into the wealth of information that can be deduced from this information about the current state-of-art of our planets.

Yet, the gravity models employed for these investigations exhibit serious deficiencies:
- Lunar gravity models suffer seriously from the lack of directly observed far side data,
- Gravity models of Moon, Mars, and Venus suffer to some extent from the spatial variations of the geometry of the connection Earth observatory to planetary orbiter and from the limited variety of orbit parameters of the orbiters. Despite the high spherical harmonic resolution of the gravity models, their actual significance does usually not exceed degree and order 20 or 30.
- Gravity information from all other planets, planetary sub-satellites and asteroids is derived from fly-byes or orbit perturbations. So far it only provides the most elementary gravity related information.

It is therefore recommended to develop dedicated gravity gradiometers for planetary missions. They should be full-tensor, nine component instruments that are capable of providing multidimensional gravity field information and are at the same time supporting positioning (orbit determination) and attitude determination.

The instrument should be robust, ambient temperature, and its precision tailored to the somewhat relaxed needs of planetary sciences. Ideally such a gradiometer should be accompanied by an altimeter, as altimetry is the second key quantity of planetology. Gravity gradiometers should simply be a standard equipment of any future lunar or planetary mission. Gradiometry is preferable to satellite-to-satellite tracking because of its compactness and because multi-spacecraft configurations add considerable complexity to planetary missions.

Acknowledgement

The contributions by Th. Peters and several important suggestions by B. Vermeersen are gratefully acknowledged.

References

Balmino, G.: 1993, *The Spectra of Topography of the Earth, Venus and Mars* AGU, p. 4.

Floberghagen, R., 2002,. *Lunar Gravimetry: Revealing the Far-Side, Astrophysics and Space Science Library*. Dordrecht: Kluwer 286.

Kaula, W. M.: 1991, in: O. L. Colombo (ed.), *From Mars to Greenland: Charting Gravity with Space and Airborne Instruments*, IAG Symposia 110, Springer, New York, 1–10.

Kiefer, W. S., Potter, E. -K.: 2000, *Lunar Planet. Sci.* **31**.

Konopliv, A. S. Binder, A. B. Hood, L. L. Kucinskas, A. B. Sjogren, W. L. and Williams, J. G.: 1998, *Science*. **281**, 1476–1480.

Konopliv, A. S. Banerdt, W. B. and Sjogren, W. L.: 1999, *Icarus*. **139**(1), 3–18.

Lemoine, F. G. Smith, D. E. Rowlands, D. D. Zuber, M. T. Neumann, G. A. Chinn, D. S. and Pavlis, D. E.: 2001, *J. Geophys. Res.*. **106**(E10), 23359–23376.

McKenzie, D. Ford, P. G. Johnson, C. Parsons, B. Sandwell, D. T. Saunders, S and Solomon, S. C.: 1992, *J. Geophys. Res.*. **97**, 13533–13544.

Nerem, R. S.: 1995, *Rev. Geophys.* Supplement, 477–480.

Rappaport, N. J. Konopliv, A. S. Kucinskas, A. B. and Ford, P. G.: 1999, *Icarus*. **139**(1), 19–31.

Schubert, G. Moore, W. B. and Sandwell, D. T.: 1994, *Icarus*. **112**, 130–146.

Schubert, G., Turcotte, D. L., and Olson, P.: 2001, *Mantle Convection in the Earth and Planets*, Cambridge University Press, p. 940.

Smith, D. E. Sjogren, W. L. Tyler, G. L. Balmino, G. Lemoine, F. G. and Konopliv, A. S.: 1999, *Science*. **286**, 94–97.

Thomas, P. C.: 1991, *Rev. Geophys.* Supplement, 182–187.

Watts, A. B.: 2001, *Isostasy and Flexure of the Lithosphere*, Cambridge University Press, 458 pp.

Zuber, M. T. Solomon, S. C. Phillips, R. J. Smith, D. E. Tyler, G. L. Aharonson, O. Balmino, G. Banerdt, W. B. Head, J. W. Johnson, C. L. Lemoine, F. G. McGovern, P. J. Neumann, G. A.Rowlands, D. D. and Zhong, S.: 2000, *Science*. **287**, 1788–1793.

Earth, Moon, and Planets (2005) 94: 113–142
DOI 10.1007/s11038-004-7605-x

SCIENCE REQUIREMENTS ON FUTURE MISSIONS AND SIMULATED MISSION SCENARIOS

NICO SNEEUW[*]

Department of Geomatics Engineering, University of Calgary, 2500 University Drive N.W., Calgary, Alberta, Canada, T2N 1N4 (E-mail: sneeuw@ucalgary.ca)

JAKOB FLURY and REINER RUMMEL

Institut für Astronomische und Physikalische Geodäsie, TU München (E-mail: flury@bv.tum.de)

(Received 8 October 2003; Accepted 11 November 2004)

Abstract. The science requirements on future gravity satellite missions, following from the previous contributions of this issue, are summarized and visualized in terms of spatial scales, temporal behaviour and accuracy. This summary serves the identification of four classes of future satellite mission of potential interest: high-altitude monitoring, satellite-to-satellite tracking, gradiometry, and formation flights. Within each class several variants are defined. The gravity recovery performance of each of these ideal missions is simulated. Despite some simplifying assumptions, these error simulations result in guidelines as to which type of mission fulfils which requirements best.

Keywords: Error simulations, gravity field, geoid, geoscience requirements, satellite missions

1. Science Requirements

Based on the analysis of the science issues and of the gravity and geoid requirements of the previous articles in this issue it is now possible to formulate a Science Requirements Table (Table I). It is organized according to the scientific areas: solid Earth geophysics, hydrology, ocean, global sea level monitoring, ice, geodesy, atmosphere, and planets, and lists the corresponding science themes. It contains estimates of the typical spatial and temporal scales and the required measurement accuracy, and gives some comments e.g. on the necessary measurement duration. It should be understood that the given numbers are only indicative. Since some of the considered phenomena have never been measured so far the estimate of their size may be off by an order of magnitude. The purpose of this Science Requirements Table is to summarize the main facts relevant for the design of future mission strategies.

For the discussion of its essence a series of so-called bubble plots has been extracted from it, see Figures 1–3. The bubble plots display all discussed science themes according to their typical spatial and temporal scales. The

[*]Current address: Institute of Geodesy, Universität Stuttgart, Germany (E-mail: sneeuw@gis.uni-stuttgart.de)

TABLE I

Synopsis of future science requirements in terms of geoid and gravity field knowledge

Science Area Theme	Required Resolution (km)	Main Periods	Required Accuracy Geoid or gravity	Comment
Solid Earth				
Glacial isostatic adjustment	> 500	10 000–100 000 y	1–10 μm/y	total geoid effect: 1–2 mm/y
Co-/post-seismic deformation, slow/silent earthquakes	regional	instantaneous—decadal	sub-mm	requires monitoring mission
Plate tectonics, mantle convection, volcanos	> 10	secular, instantaneous	< 1 mm/y	requires monitoring mission
Core motion (nutation, Slichter), seismic normal modes	> 5000	10 s–18 y	1 nGal – 1 μGal	
Hydrology				
Snow, precipitation, ground water, dams, soil moisture, run-off, evapo-transpiration	10–5000	1 h–secular	0.5–1 mm monthly	high spatial resolution more important than accuracy
Ocean				
Mean flow: narrow currents, topographic control	20–50	quasi-static	5–10 mm	
Coastal currents along shelf edges	10–50	quasi-static	5–10 mm	
Interaction mean and eddy flow, ocean fronts position	10–100	quasi-static	5–10 mm	
Bathymetry	1–10	static		

Basin scale mass change, deep water formation	1000–5000	months – decades	10 mm	sea level, oscillations
Bottom currents	10–200	months – decades	0.1–1 mm	
Sea Level				
Global sea level change monitoring	> 2000	interannual, secular	0.1 mm/y	monitoring mission desired
Ice				
Ice mass balance	100–4000	seasonal – secular	< 0.01 mm/y	monitoring mission desired
Bottom topography, ice compaction	20–50	quasi-static	0.01–0.1 mGal	
Geoid for sea ice thickness	10–100	static	100 mm	
Geodesy				
Precise heights for engineering, GNSS levelling, coastal height reference, sea level monitoring	20–50	static	5–20 mm	for some areas also geoid time variation
Inertial navigation	5–10	static	0.1 mGal, 0"1 deflection of vertical	combination with terrestrial data required
Atmosphere	gravity field improvement may be interesting for future atmospheric modeling			
Planets	dedicated autonomous gravity field missions with very high spatial resolution			

spatial scales are arranged logarithmically along the horizontal axis, the temporal scales are ordered along the vertical axis. The scales should give only an approximate indication.

Figure 1 displays all themes according to their field of Earth science. Top priorities are indicated by ellipses with thick borders. The two diagonally shaded bubbles are associated to phenomena (atmosphere and tides) that are not considered science priorities but that are contained in the measurement signal and need to be isolated.

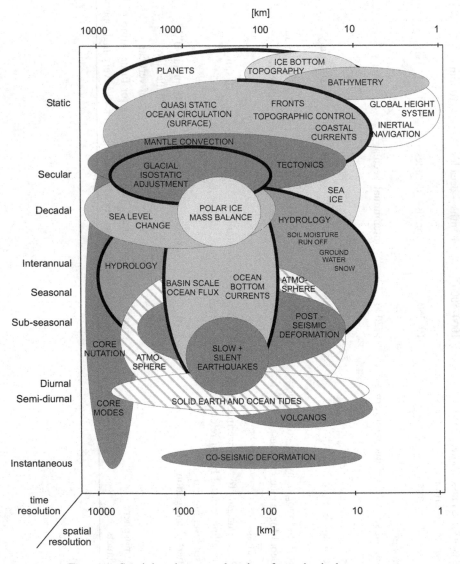

Figure 1. Spatial and temporal scales of geophysical processes.

Figure 2 contains frames indicating the spatial and temporal scales covered by the GRACE (JPL, 1999) and GOCE (ESA, 1999) satellite missions. The logarithmic scale associated with the spatial variations is thereby somewhat misleading, as discussed in (Balmino et al., 1998). Nevertheless the picture tells us that future missions can improve the state-of-art after GRACE and GOCE in three ways:

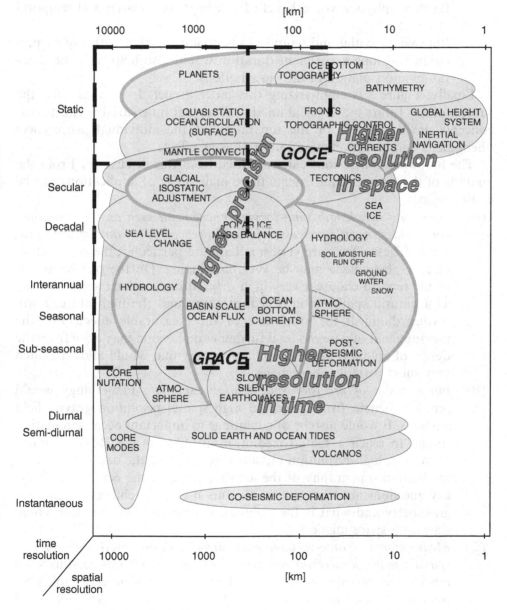

Figure 2. Gravity field recovery requirements after GRACE and GOCE.

(i) Achieving a higher precision – without necessarily improving the spatial or temporal resolution – and therefore improving the quantification of the phenomena already measured by GRACE and GOCE.

(ii) Increasing the spatial and/or temporal resolution: The study of coastal currents, fronts, ice bottom topography, bathymetry, inertial navigation and height determination would primarily benefit from an improved spatial resolution, while hydrology, deep ocean studies, solid Earth geophysics would benefit from improved spatial and temporal resolution.

(iii) Improving spatial and temporal resolution together with higher precision and long experiment duration would also help with the separation and isolation of individual effects.

Finally, Figure 3 is emphasizing this latter aspect. It is crucial for the success of any future gravity field mission aiming at time variable effects that concepts are developed for the separation of the individual geophysical phenomena and for the elimination of aliasing effects.

The following conclusions can be drawn from this discussion. From the analysis of the science requirements three main areas of applications can be distinguished:

(A) *Very long wavelength time-varying phenomena such as core nutation, core modes, mantle processes or secular oceanic or atmospheric processes*: These effects are very small and, in particular, the observation of core phenomena may be very challenging. On the one hand one would tend to choose a very high orbit altitude, which acts so-to-say as a natural spatial filter. On the other hand the high altitude will further diminish the amplitude of the measurable effect. For the measurement of core modes a further complication may arise from the design of an appropriate sampling strategy that would allow to catch very short periodic phenomena.

(B) *Improvement of precision and spatial resolution*: Planetology would certainly benefit from dedicated high spatial resolution gravity field missions. It would also be of advantage to important ocean circulation studies to reach a spatial resolution as short as the Rossby radius. Moreover, high-resolution missions would serve the determination of the bottom topography of the ice sheets and of the oceans. Here the key questions are what spatial resolution can be achieved by satellite gravimetry and what is the ultimate resolution limit that still would warrant a space mission.

(C) *Measurement of time-variable phenomena with improved precision and spatial and temporal resolution*: It can be foreseen that a successful GRACE mission will prompt the study of a large number of time-variable phenomena in the area of solid Earth geophysics, oceanography and

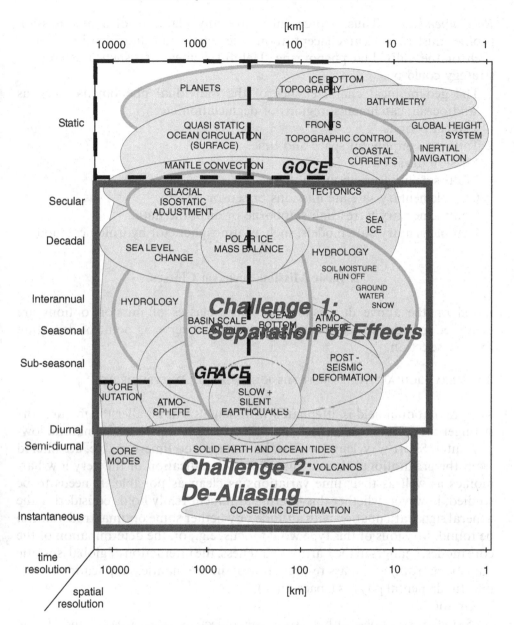

Figure 3. Challenges for future gravity satellite missions.

hydrology, pushing for ever higher precision and resolution. The complication is, however, that the scientific analysis requires each of the phenomena to be quantified individually while only their integral effect can be measured. In addition, some of the interesting time-varying phenomena are rather small and often superimposed by another, much larger process in its space–time window.

Recommendation: Thus, a prerequisite for any planning of a new mission profile must be a clear concept about the separation and identification of each of the individual phenomena. Helpful elements to such a separation strategy could be:
- The geographical characteristics of the individual phenomena, such as land/ocean, catchments, regions of deglaciation etc.
- Typical periods.
- Tailored sampling in space and time.
- Mission duration.
- Multi-satellite formation flight.
- Complementary satellite missions or sensor systems.
- Complementary terrestrial, shipborne or airborne data.
- Complementary data models, such as atmospheric or hydrological models.

2. Basic Mission Scenario Classes

Based on the above discussion four basic classes of mission options are identified in order to set the scene. They should serve as point of departure for any selection.

2.1. VERY HIGH ALTITUDE ORBIT MISSIONS (LAGEOS TYPE ORBITS)

The gravitational field is attenuated with height. The smaller the features the stronger the attenuation. A high altitude orbit acts therefore as a natural low-pass filter. Short-wavelength effects, both static and time-variable, are filtered from the gravitational signal. This allows identification of the very low harmonics as well as their time variations as clean as possible. It needs to be studied, however, whether this filtering argument really holds considering the general signal attenuation with altitude or whether some optimal trade-off can be found. Missions of this type would focus, e.g., on the determination of the core modes, changes in the Earth's oblateness, the interaction of global isostatic adjustment (GIA) and mass re-organization in the mantle. Applications reach into fundamental physics (space–time).

Options:
(1) Satellites equipped with passive laser reflectors, like LAGEOS I and II (e.g. http://www.earth.nasa.gov/history/lageos/lageos.html).
    - Advantages are long mission life time, simplicity and low cost.
    - A disadvantage is the lack of continuous and three dimensional tracking.
(2) Compact satellites equipped for high-low tracking to a Global Navigation Satellite System (GNSS), additionally equipped with an accelerometer.

- Advantage is the three dimensional and uninterrupted tracking. One has to study, however, whether the orbits/tracking based upon a GNSS is stable and accurate enough (is GNSS an accurate enough reference?)
- A severe disadvantage could be the limited life time due to the electronic equipment.
- A life time of 10 yrs and more would be desirable.
(3) A single satellite or several satellites furnished with GNNS receivers, accelerometers, as well as high precision clocks (10^{-16}).
  - This would be a new approach using the general relativistic influence of the position dependent gravitational potential on time.
  - Such a mission could be of interest for Earth sciences, time keeping, fundamental physics and telecommunication. The latter point needs further clarification.
  - A fundamental shortcoming could be the limited life time.

2.2. FUTURE GRADIOMETER MISSIONS

Gradiometry should be considered the prime candidate for any planetary mission. The complexity of a single-satellite mission seems preferable over a multi-satellite planetary Satellite-to-Satellite Tracking (SST) mission, despite the complexity of the gradiometer instrument itself. Moreover, a full tensor system, i.e. a gradiometer measuring all nine possible components, could serve as a combined gravimeter–navigator. The navigation part would support orbit recovery (aided by high–low tracking from Earth) and attitude determination (aided by star sensing).

Gradiometry could also be used for further improvement of the Earth's static field.

For the future of gradiometry the following basic questions need to be addressed:
- How much technological improvement is possible beyond the current gradiometry accuracy level of about $10^{-3}E/\sqrt{Hz}$?
- In order to improve spatial resolution what is the lowest possible orbit?
- What is the longest possible mission life time?
- What could be a logical instrument development line (from ambient temperature, via superconducting to quantum gradiometers; rotating versus non-rotating, free floating test masses)?
- For a planetary gradiometric mission: What is the optimal trade-off between orbital height and gradiometer sensitivity?
- How do the above points affect the complexity of total sensor concept, the choice of the material, the propulsion system etc.?

2.3. FUTURE LOW–LOW SST MISSIONS

Generally, a next generation low–low SST mission is considered a prime candidate for refined measurement of temporal variations in Earth system components. The main argument for that is the alleged precision of the inter-satellite distance measurement. But what are the technological limits of the system as a whole? What are the consequences of non-isotropic spatial correlation of the gravity field sensed in one direction?

Basic questions to be addressed are:

- If a lower mission altitude is desirable how does it add to the complexity of the mission?
- If the precision of inter-satellite tracking (laser or microwave) is more and more improving what are the corresponding requirements for accelerometer and star sensor systems? Are they the limiting factor?
- Related to the above question: What is the signal spectral behavior of the non-gravitational forces (drag, radiation pressure, magnetic field etc.) to be measured by the accelerometer system? "If there is no high-frequency disturbance it does not have to be measured."
- What is the optimal inter-satellite tracking concept? Also in view of the mission life time.
- How difficult will it be to separate the various geophysical signals? This is probably the most fundamental question in general, but even more when discussing a one component instrument? See the recommendations in Section 1.

2.4. CONFIGURATION FLIGHTS

Configuration flights seem to get fashionable in space science. What can they do for Earth sciences? Their basic purpose here is to measure the Earth's gravity field by inter-satellite tracking simultaneously in several spatial directions (like it is done in gradiometry with a single instrument). So, the first question is, (a) why it is desirable to measure several spatial directions and (b) why simultaneously. It is not trivial to answer these questions in a profound manner. The issues are separation of geophysical effects and de-aliasing. Any gravity measuring device is only able to measure the lumped sum of all (time-variable) gravitational signals. In addition, some of the "high-frequency" time-varying signals (diurnal, semi-diurnal) alias into time-variable signals of much longer periods.

It is nevertheless the scientific goal to be able to study the individual contributions separately. How can this be done? Partly, the phenomena are confined to certain geographical regions (land, sea, ice, plate boundaries) and time scales. This may help. Also, using known models about the space–time structure of some of the geophysical signals may prove helpful. Moreover, complementary measurements (terrestrial, airborne, space missions) can and should be employed.

The measurement of independent gravity field components may prove essential. It is like seeing the same effects from various directions. This should help to discern certain classes of signal and also to de-alias signals. A thorough study of this aspect is of fundamental importance for the justification of any future mission that aims at measuring time variable gravity.

Configuration flights of several satellites (such as a Cartwheel or LISA configuration) can be established by means of slight differences in inclination, ascending node and eccentricity of the orbits of the participating satellites. A configuration can also be maintained or supported actively by means of thruster systems.

The main gravity field observable of a configuration would be SST in the low–low mode. However, depending on the configuration, this would take place in several directions. Moreover, these directions are time-variable. Alternatively, one could interpret such a measurement configuration as gravity gradiometry.

3. Mission Performance Simulations

3.1. PRELIMINARY REMARKS

Spherical harmonic error analysis: The word *(error) simulation* here means simulation of a spherical harmonic error spectrum, that may be obtained from a given gravity mission scenario. The technique is well-documented, cf. (Schrama, 1991) or (Sneeuw, 2000).

A number of simplifying assumptions is used in such error simulations, e.g. uninterrupted data stream over the mission length, or linearization of the observation model using a nominal orbit of constant height and inclination. It has been demonstrated, though, that in spite of these simplifications the simulated error spectra are representative of more realistic scenarios. The simulation results in this report can safely be regarded as realistic estimates of the gravity field error level from the given mission scenarios.

At the same time one should be aware that the Earth's gravity field in these error simulations is implicitly a *static* field. A gravity mission that aims at measuring gravity field variations involves many more issues related to the intricate sampling in time and space: ground-track variability, aliasing, snapshot solutions. These issues cannot be captured by the current simulation software. All output described in this report must be understood as a representation of the basic gravity field sensitivity of a certain mission scenarios.

Whether this sensitivity can be used to the full extent for time-varying gravity field recovery remains to be analysed by other means.

One of the stronger assumptions in these simulations is that the observable is a purely gravitational one. In other words, it is assumed that all subsystems (accelerometers, drag-free control, attitude control) perform at least as well or are calibrated at the same level as the indicated measurement precisions.

Graphical representation: The basic output of an error simulation is a block-diagonal error covariance matrix of the spherical harmonic coefficients. These covariance matrices are not suitable for direct graphical representation.

The simulated spherical harmonic error spectra will be visualized in a condensed way, using three panels. The left panel will show RMS curves per degree. An error degree RMS (RMS_l) is defined as:

$$RMS_l = \sqrt{\frac{1}{2l+1} \sum_{m=0}^{l} \left(\sigma_{\bar{C}_{lm}}^2 + \sigma_{\bar{S}_{lm}}^2 \right)},$$

where $\sigma_{\bar{C}_{lm}}^2$ is the error variance of the spherical harmonic coefficient $\bar{C}_{lm}$. It is taken from the main diagonal of the aforementioned error covariance matrix. As opposed to the more conventional *degree variance*, a *degree* RMS represents the magnitude of either error or signal of a single coefficient. For comparison purposes two other degree RMS curves will be displayed: a signal curve as implied by Kaula's rule and the error level from EIGEN2. The latter represents state-of-the-art gravity field modelling from CHAMP data, cf. (Reigber et al., 2003). All spectra in the left panels are dimensionless.

The middle panels represent the *geoid commission error*. It is a cumulative error, lumping the full error variances from degree 2 to a given degree. Error levels are in units of metres.

Similarly, the right panels represent the *gravity commission error*, expressed in mGal.

3.2. MISSION SCENARIO DEFINITIONS

In each of the four mission scenario groups discussed in Section 2 certain variants are defined in terms of the orbital characteristics, observation types and measurement precision (also refer to Table II):

1. *Very high altitude monitoring missions:* One based on GNSS tracking, and one based on low–low SST using laser technology. Orbit altitude is 6000 km (LAGEOS–like). The inter-satellite distance (ISD or ρ_0) is an arbitrary 1000 km, see also the discussion in Section 4.3.2.

TABLE II

Mission scenarios and parameters. Δx stands for 3D orbit perturbations (GNSS) and $\Delta\rho$ stands for low–low SST range measurements. ρ_0 is the nominal inter-satellite distance. The maximum spherical harmonic degree of the simulation is designated as L. In general it does not correspond to the degree of resolution

Numbers of	Observable	PSD (unit/$\sqrt{Hz}$)	h (km)	L	ρ_0 (km)	Comment
1. Very High Altitude Monitoring Missions						
1a	Δx	1 cm	6000	30	–	hi–lo SST at LAGEOS altitude
1b	$\Delta\rho$	0.1 μm	6000	30	1000	lo–lo SST at LAGEOS altitude
2. Gradiometry Missions						
2a	T_{xx}, T_{yy}, T_{zz}	1 mE	180	400	–	
2b	T_{xx}, T_{yy}, T_{zz}	0.1 mE	180	600	–	
2c	T_{xx}, T_{yy}, T_{zz}	1 mE	100	600	–	represents high e mission
2d	T_{xx}, T_{yy}, T_{zz}	0.1 mE	100	720	–	
2e	full tensor	10 mE	50	600	–	planetary mission
3. Low–Low SST Missions						
3a	$\Delta\rho$	10 μm	400	150	50	current GRACE, short ρ_0
3b	$\Delta\rho$	10 μm	200	300	50	very low GRACE
3c	$\Delta\rho$	0.1 μm	400	300	50	3a 100× better (laser)
3d	$\Delta\rho$	0.1 μm	200	500	50	3b 100× better (laser)
3e	$\Delta\rho$	0.1 μm	200	500	200	3d with larger baseline
3f	$\Delta\rho$	0.1 μm	200	500	1000	3d with very large baseline
4. Configuration Flights						
4a	T_{xx}, T_{xz}, T_{zz}	10 μE	400	400	0.1–1	CartWheel-like (coplanar)
4b	full tensor	10	400	400	0.1–1	LISA-like (non-coplanar)

2. *Gravity gradiometry missions:* Two different accuracies (1 vs. 0.1 mE/$\sqrt{Hz}$) are tested at two orbit altitudes (180 vs. 100 km). The former altitude may be achieved with a high-quality drag-free control (DFC) system. The latter height is not sustainable. A mission of this type should be realized through an eccentric orbit, see Section 4.2.2. Additionally, one scenario at ultra-low altitude (50 km) simulates a lunar or planetary mission using a lower quality (10 mE/$\sqrt{Hz}$) full tensor gradiometer.

3. *Low–low SST missions:* Current metrology (based on radio frequency technology) is tested at GRACE altitude (400 km) and at a very low altitude (200 km). The latter would require an active DFC. In both cases a short baseline (50 km) is assumed. At the same two altitudes and with

the same baseline length, next generation metrology is tested assuming a PSD of 0.1 $\mu m/\sqrt{Hz}$, which may either be realized through laser interferometry (Bender et al., 2003) or atomic interference (McGuirk et al., 2002). The high-precision low-orbit scenario is repeated with two different baselines (200 and 1000 km) to test common-mode attenuation effects, cf. Section 4.3.2.

4. *Configuration flights:* To test the feasibility of configuration flights for gravity field mapping purposes, two configurations are tested. One CartWheel-like constellation, in which all satellites fly coplanar, and one LISA-like constellation, in which the satellites perform a relative circular motion which requires an out-of-plane motion as well. Both constellations are simulated at an average height of 400 km. It is assumed that the observable from such constellations can be converted into gravity gradients. See also Section 4.

The following assumptions and parameters have been used for the simulations:
- All heights are heights above the equator.
- All orbits are polar ($I = 90°$).
- All missions last 1 yr.
- All measurement error power spectral densities (PSD) are assumed to be white.

3.3. SIMULATION PARAMETER EFFECTS

The following rules of thumb can be applied to the simulation results, if other mission parameters would be considered. The effects mentioned below are very similar to the basic mission design options, described in (Rummel, 2003). In general, these rules are valid when no regularization takes place and when the orbit is polar.

Instrument accuracy: The instrument accuracy, expressed as power spectral density in Table II scales linearly into the output RMS_l and the geoid and gravity commission errors. Roughly, a 10 times better instrument yields a 10 times better gravity field.

Mission duration: All simulations are based on a mission duration of 1 yr for purposes of comparability. However, missions like monitoring missions (Category 1) would last a lot longer. Mission duration scales into the error outputs using the square root. Thus a mission of 4 yrs would provide a twice better result. Care must be taken with this rule of thumb. It only applies to the static gravity field as a single result of the whole mission. Monitoring missions with time-variable gravity as a mission objective, will provide gravity in terms of snap-shots in time. In such cases one cannot scale over the full mission duration. One should scale over the duration of the snapshot, instead. A monthly snapshot would be a factor $\sqrt{12}$ worse than the presented simulation result.

Orbit height: The gravity potential and its functionals attenuate with height by $(R/r)^{l+i}, i = 1, 2$ or 3. A lower orbit means a higher sensitivity towards the higher degrees. The effect of lowering an orbit on error degree spectra is a rotation of the RMS$_l$-curve, i.e. a change of slope, around a point at the left of the curve.

Inclination: For all mission scenarios a polar orbit ($I = 90°$) is assumed. The resulting error levels are representative for non-polar orbits (e.g. sun-synchronous) as well, although they would be applicable to the areas covered by ground-tracks only. Around the poles, circular data gaps would exist, where the simulated error levels are invalid.

4. Simulation Results

The simulated error spectra for all variants will be displayed and commented in the following section. Figures are annotated with numbers that represent a certain mission scenario. These mission codes refer to the 1st column of Table II.

4.1. HIGH-ALTITUDE MONITORING MISSIONS

Scenario 1a represents a simple satellite configuration, carrying a GNSS receiver as main instrument. The RMS$_l$ curve in Figure 4 shows that such a mission is hardly of scientific use. The very low degrees are recovered with a precision that is similar to the current CHAMP solution EIGEN2.

A high altitude monitoring mission, aiming at the recovery of the very low degrees, should have a high-precision tracking system. Scenario 1b simulates a GRACE-like mission at LAGEOS altitude using a laser-tracking ($0.1 \mu m/\sqrt{Hz}$) over a 1000 km baseline. Figure 4 reveals the extremely high sensitivity towards the low degrees. The cumulative geoid error up to $l = 10$ is less than a μm.

Figure 4 also shows the steep ascent of all error spectra. This means that the high altitude works as an efficient low-pass filter. The higher degrees are filtered out and cannot contaminate the lower degrees. For Earth core research an even higher orbit may be considered.

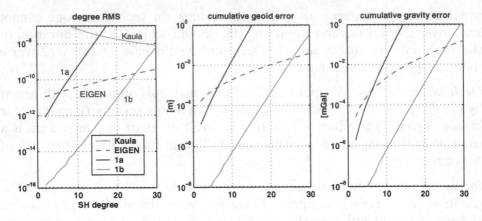

Figure 4. Error spectra of the high altitude scenarios 1a and 1b.

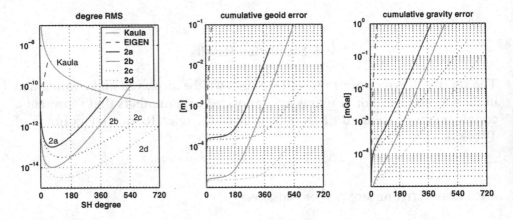

Figure 5. Error spectra of the gradiometric scenarios 2a–2d.

4.2. GRAVITY GRADIOMETRY

A number of error simulations have been run to investigate post-GOCE gradiometer missions. Results are shown in Figure 5.

4.2.1. Orbit height vs. instrument accuracy

The two basic variables for the gravity gradiometry simulations are orbit height and instrument accuracy. Assuming a high-quality DFC system, an orbit height of 180 km seems feasible. This is simulated in scenarios 2a and 2b.

It will be hard to fly lower than that, at least if the orbit height has to remain constant. To decrease the altitude, the idea of an eccentric orbit was brought up. A perigee height of 100 km was chosen. The current simulation software is not able to handle eccentric orbits. The simulations 2c and 2d have been run, as if the satellite was at a constant altitude. The issues with eccentric orbits concerning sampling rate, zoom-in effect, or perigee precession are discussed in Section 4.2.2.

Two gradiometer accuracies have been tested. The 1-mE gradiometer represents current gradiometer technology (missions 2a and 2c). The 0.1-mE gradiometer represents future gradiometer technology (2b and 2d).

In all four cases, a GOCE-like gradiometer was assumed. Only the in-line components T_{xx}, T_{yy}, T_{zz} are measured.

A GOCE-like mission at a lower orbit (2a) achieves a solution up to degree $L = 420$, i.e. 48 km spatial resolution. With a better instrument (2b) $L = 500$ or 40 km spatial resolution is achieved. At these maximum degrees, the geoid commission errors are 5 and 3.5 cm, respectively.

Mission scenarios 2c and 2d, representing eccentric orbits, achieve far higher resolutions: 28 km ($L = 720$) and 22 km ($L = 900$) for the 1-mE and the 0.1-mE gradiometers, respectively. If we extend the error curves for 2c and 2d in Figure 5, we can read the geoid commission error at these resolutions. For both scenarios the geoid commission error is around 1.5 cm.

The simulation parameter effects, described in section 3.3 are clear from Figure 5. A tenfold improvement in instrument performance reduces the error spectrum by an order of magnitude. Reducing the height, on the other hand, changes the slope of the error curve in favour of higher resolution.

4.2.2. Eccentric orbits, resolution and sampling

Perigee precession: Unless the inclination is critical ($I \approx 63.5°$), the perigee is rotating in the orbit plane. Therefore, the area of highest spatial resolution is latitude dependent. The perigee rotates due to Earth's flattening according to:

$$\dot{\omega} = \frac{3n C_{20} R_E^2}{4(1 - e^2)^2 a^2} \left(1 - 5\cos^2 I\right).$$

For a polar orbit with 100 km perigee height (above a reference Earth radius $R_E = 6378\,137$ m) and 1000 km apogee height, this amounts to $\dot{\omega} = 3.8°/\text{day}$. In this case we would have a semi-major axis $a = 6928\,137$ km and an eccentricity $e = 0.065$. The perigee would perform one revolution in nearly 100 days.

At critical inclination, the perigee is stationary at a chosen latitude. A mission that uses such an orbit will therefore be able to *zoom in* on the gravity field in a certain latitude band.

On the other hand, perigee precession can be used to assure that the full Earth has been covered by maximum resolution scanning. With the aforementioned orbit parameters as an example, every latitude band has been scanned twice in 100 days. This is approximately the right precession rate – and latitude scanning rate – for a scenario like 2d that may achieve $L = 900$ resolution. A rule of thumb ($\beta > 2L$) says that at least 1800 revolutions will be needed for scenario 2d. One orbit revolution takes about 1.6 h. Thus, nearly 120 days of data would be needed to accomplish this mission.

A larger perigee precession rate requires a lower a and a larger e. If a fixed perigee height is a given parameter, then the following formulas can be used to relate apogee/perigee height to a and e:

$$\left. \begin{array}{l} h_p = a - ae - R_E \\ h_a = a + ae - R_E \end{array} \right\} \Leftrightarrow \left\{ \begin{array}{l} a = R_E + \frac{1}{2}(h_a + h_p) \\ e = \frac{h_a - h_p}{2R_E + h_a + h_p}. \end{array} \right.$$

Variable sampling rate: From the *vis-viva* equation, giving the total energy of a Kepler orbit, one can derive the relation between linear velocity, semi-major axis and eccentricity:

$$\frac{v^2}{2} - \frac{GM}{r} = -\frac{GM}{2a} \quad \Rightarrow \quad v^2 = GM\left(\frac{2}{r} - \frac{1}{a}\right).$$

For perigee (p) and apogee (a) we have:

$$r_p = a(1 - e) \Rightarrow v_p^2 = \frac{GM}{a}\frac{1 - e}{1 + e},$$
$$r_a = a(1 + e) \Rightarrow v_a^2 = \frac{GM}{a}\frac{1 + e}{1 - e}.$$

With the assumed orbital parameters, the linear velocity varies from 7.1 km/s at apogee to 8.1 km/s at perigee. Thus, a constant sampling rate in time translates into a variable spatial sampling rate. At perigee the spatial sampling distance would be maximum. However, since the sensitivity is maximum at perigee, too, the sampling rate should be highest there.

Mission scenario 2d, for instance, has a resolution close to $L = 900$, translating into spatial scales of about 40 km full wavelength. In 5 s the satellite overflies one such minimum wavelength. The sampling rate must therefore be 0.4 Hz at least. Moving away from perigee yields a relaxed sampling. Beyond a certain orbital height, when sensitivity degrades, one could even stop measuring.

4.2.3. Lunar or planetary missions
Figure 6 compares the results of scenario 2d (0.1 mE-gradiometer at 100 km height through eccentric orbit) to scenario 2e, which assumes a lower quality

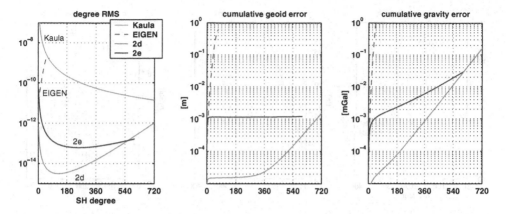

Figure 6. Error spectrum of the planetary scenario 2e in comparison to 2d.

full-tensor gradiometer (10 mE/$\sqrt{\text{Hz}}$) on an ultra-low (50 km) orbit. Scenario 2e simulates a lunar or planetary mission, in which the lack of an atmosphere allows to fly this low.

For comparison's sake the mission has been simulated with the Earth's parameters, e.g. mass and size. The results will be indicative nevertheless for Mars or Moon, whose parameters are of the same order of magnitude as the Earth. In particular, the resolution should be expressed in angular measure, i.e. independent of planetary size.

From Figure 6 it can be seen that a lower-quality gradiometer is able to outperform a high-quality instrument, if the orbit is low enough. At the very high degrees, beyond $l = 560$, mission 2e even performs better than 2d. The resolution would be around $L = 1200$ (0.15°). The cumulative geoid error of 2e in the middle panel of Figure 6 may not be entirely representative for Moon or Mars. The important characteristic, however, is the flatness of the cumulative curve over a wide frequency band. The flatness is caused by the low orbit. The actual level depends on the instrument precision.

4.3. SST MISSIONS

4.3.1. Orbit height vs. instrument accuracy
Again the two variables height and accuracy are tested (See Figure 7). Mission scenarios 3a and 3c represent low–low SST missions at 400 km height, whereas 3b and 3d simulate a very low orbit (200 km), which would require DFC. Missions 3a and 3b share the PSD of 10μm/$\sqrt{\text{Hz}}$, representing current radio frequency technology. Missions 3c and 3c simulate a 100 times better laser-based SST. All missions assume a 50 km baseline.

Mission 3a produces a RMS$_l$ similar to the expected GRACE results. The resolution is close to 133 km ($L = 150$). At this resolution, the geoid

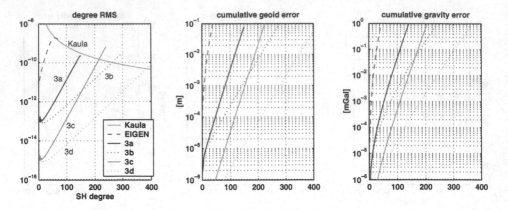

Figure 7. Error spectra of the low–low SST scenarios 3a–3d.

commission error becomes approximately 10 cm. If the same configuration would be flown lower (mission 3b), it could achieve a solution up to $L = 260$, close to that of GOCE, also with a geoid commission error of nearly 10 cm.

The RMS$_l$-curves of 3c and 3d show the same behaviour, though two orders of magnitude lower. Mission 3c results in a resolution of 91 km ($L = 220$) where the geoid commission error is a little bit less than 10 cm. Mission 3d, representing laser-SST at a very low orbit, achieves a resolution of 51 km ($L = 390$) with a corresponding geoid commission error of 4 cm. The latter result is very similar to the high-quality gradiometry mission 2b.

4.3.2. Baseline length and common-mode effects

As a rule the baseline between the satellites should be shorter than the minimum spatial scales to be resolved. If the baseline is too long, common-mode attenuation may occur. At certain wavelengths, both satellites would go through exactly the same orbit perturbations, in which circumstance the ISD would be insensitive to gravity field features of this wavelength.

The attenuation behaviour is clear from Figure 8. Mission 3d with a 50 km baseline still has a smooth error spectrum. One attenuation peak is visible in the RMS$_l$-curve of mission 3e, which has a 200 km baseline. A second attenuation peak would occur at the right-hand edge of the graph at degree 400. Mission 3f with a baseline of 1000 km has a reduced accuracy every 40 degrees in the error spectra.

Mathematically, the SST observation equation contains a term $\sin(\eta \beta_{mk})$ in which η is half the satellite separation in angular measure, and β_{mk} the normalized frequency, which has units of cycles-per-revolution. One can approximate β_{mk} by k, the along-orbit wavenumber. The spherical harmonic degree l is always larger than $|k|$. Thus, attenuation occurs when k is close to $i\pi/\eta, i = 0, 1, 2, \ldots$.

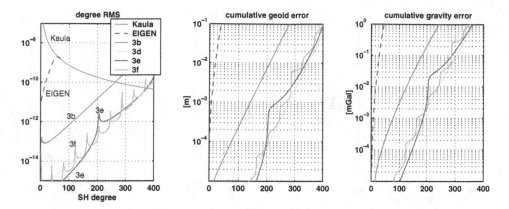

Figure 8. Error spectra of the low–low SST scenarios 3d–3f. Effect of variable baseline

For instance, a 1000-km baseline comes down to $\eta = 4.5°$. Indeed, $180°/4.5°$ equals 40. For mission 3e, η would be $0.9°$ leading to $180°/0.9° = 200$, which explains the attenuation peak for 3e. The baseline choice of 50 km for all other SST-missions would have its first attenuation at $l = 800$.

Despite the attenuation bands, the cumulative geoid and gravity error curves of missions 3e and 3f remain relatively close to the spectra of 3d. Nevertheless, a non-homogeneous error spectrum should be avoided for any gravity mission. A simple matching of baseline length to the required spatial resolution is sufficient for that purpose.

4.4. GRADIOMETRY FROM CONFIGURATION FLIGHTS

From (Bender et al., 2003), discussing heterodyne laser interferometry, or (McGuirk et al., 2002), discussing atomic interference, differential accelerometry seems feasible at a level of $10^{-12} m/s^2/\sqrt{Hz}$. Over a baseline of 100 m this would translate into gradiometry at the $10^{-5} E/\sqrt{Hz}$ level. Mission scenarios 4a and 4b simulate such SST-based gradiometry for two different satellite configurations at 400 km altitude (See Figure 9).

Coplanar satellites (CartWheel-type): A large body of literature exists on the topic of formation flying, e.g. (Alfriend and Schaub, 2003). The concept has been studied for several missions, e.g. LISA, DARWIN, ORION, and seems to be feasible despite J_2 effects, eccentric orbits and non-linearities.

Let us assume a simplified case in which a number of satellites, all in the same orbital plane, perform a relative elliptical motion. The relative ellipse's major axis is in along-track direction and is twice as large as the radially oriented minor axis.

For the following discussion, let x be the along-track, y the cross-track, and z the radial direction. Also, let us consider one satellite pair only. We take the

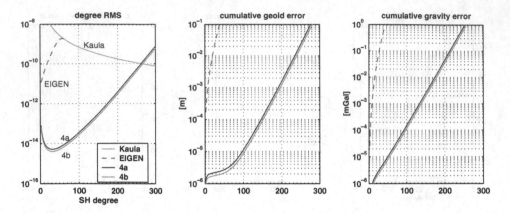

Figure 9. Error spectra of the gradiometric configuration flight scenarios 4a and 4b.

relative acceleration between the satellites as our observable. If this would be done for GRACE, we would have the V_{xx} along-track in-line gravity gradient.

The gravity gradient tensor $\mathbf{V} = V_{ij}$ transforms under a rotation of the coordinate frame as RVR^{T}, in which R denotes the rotation matrix. Now assume a rotation α about the y-axis only, such that the two satellites are always on the new x'-axis. The in-plane gradients V_{xx}, V_{xz}, V_{zz} project onto the observable as follows:

$$V_{x'x'} = \cos^2 \alpha V_{xx} + 2\cos\alpha \sin\alpha V_{xz} + \sin^2 \alpha V_{zz}.$$

This discussion assumes so far that inertial differential accelerations are measured and compensated up to the same level accuracy.

Gravity field recovery is feasible using a single satellite pair, either in GRACE-like orientation or in a CartWheel-like relative elliptical motion. The disadvantage of GRACE is the along-track in-line component that has a non-isotropic sensitivity. GRACE is basically sensitive to East-West features in the gravity field only. Radial SST would be superior to the GRACE observable. This can be achieved only partially by a single CartWheel pair. The disadvantage of a single CartWheel pair, though, is that the orientation is relatively constant in inertial space. Thus, in Earth-fixed coordinates, the sensitivity will become strongly latitude dependent.

These drawbacks are avoided by tracking three intersatellite distances. With three different angles α we would have three simultaneous equations of the above kind, leading to an instantaneous determination of V_{xx}, V_{xz} and V_{zz}. The three ISDs can either be realized by a CartWheel of three satellites, measuring in a triangle, or by six satellites, measuring along the spokes of the wheel, see Figure 10.

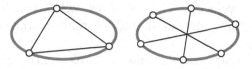

Figure 10. Potential coplanar configurations for measuring the in-plane T_{xx}, T_{xz}, T_{zz} simultaneously: triangle edges (left) or spokes (right).

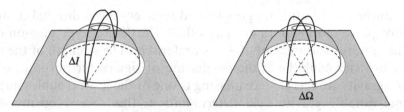

Figure 11. Non-coplanar orbits: out-of-plane components through differential inclination (left) or differential ascending node (right).

Non-coplanar satellites (LISA-type): Gradiometry of out-of-plane components (V_{xy}, V_{yy}, V_{yz}) can only be achieved through non-coplanar satellite configurations. In the simplest case, two satellites fly in different orbits that are separated in inclination, in right ascension of the ascending node, or a combination of both. In case of a differential inclination, the maximum cross-track component occurs at the pole. In case of differential ascending node, the maximum cross-track separation occurs at the equator, see Figure 11.

The in-plane motion will always be a 2:1 relative ellipse. By choosing a proper cross-track separation, the relative orbit can be made circular, as is done in a heliocentric setting for LISA, such that the ISDs remain constant. Alternatively one may choose to make the projection of the relative motion on the surface of the Earth circular.

Because of the Earth's flattening, a differential inclination is an unfortunate choice for achieving out-of-plane sensitivity. The right precession of the ascending node and the orbital acceleration due to the Earth's dynamic flattening are described by:

$$\dot{\Omega} = \frac{3nC_{20}a_E^2}{2(1-e^2)^2 a^2} \cos I,$$

$$\dot{M} = n - \frac{3nC_{20}a_E^2}{4(1-e^2)^{3/2} a^2} \left(3\cos^2 I - 1 \right).$$

Thus, a difference in inclination between two or more satellites immediately results in differential precession of the ascending node and in differential orbital velocity. The satellites will consequently drift apart. Therefore, out-of-plane motion should be realized through a difference in ascending node.

The results of missions 4a and 4b are nearly equal, cf. Figure 9. Because of the high orbit, the degree of resolution is relatively low, despite the very low gradiometric PSD of $10 \ \mu E/\sqrt{Hz}$. With $L = 270$, the spatial resolution is 74 km with a corresponding geoid commission error of 6 cm.

For purposes of stationary gravity field recovery the additional complexity of out-of-plane motion does not pay off. A CartWheel-type mission design, that guarantees measurement of T_{zz} is preferable. The strength of the out-of-plane sensitivity could be in the de-aliasing of signals for purposes of time-variable gravity recovery. Since aliasing is one of the most troublesome issues for time-variable gravity field determination, this should be investigated closely before discarding a LISA-like formation.

5. Simulation Results and Science Requirements

Table III summarizes the resolution and the accuracies for all simulation variants. In this section the simulation results from the previous sections are confronted with the accuracy requirements summarized in Table I. The aim is to demonstrate, which accuracy requirements can be met for which spatial resolutions by the mission concepts under investigation.

From the error curves for each scenario given in Figures 4–9 it can be extracted at which degree the cumulative geoid errors exceed certain accuracy levels. We extracted the maximum degrees for three geoid accuracy levels (gravity errors were not considered for this comparison):

- The 1 μm level, which would meet the strictest requirements discussed in this report.
- The 0.1 mm level, which would allow to resolve some, but by far not all of the time-variable gravity signals.
- The maximum resolution, where the signal to noise ratio is 1, with an accuracy level of 1–10 cm for most mission scenarios, these accuracies being not interesting for time-varying signals but for the static gravity field.

The maximum degrees L for these levels were transformed to half wavelengths $\lambda_{max}/2 = 20\ 000$ km/L. For four selected mission scenarios the resolutions for the three accuracy levels have been added to the *bubble plot* for the spatial and temporal scales of geophysical signals as light, medium and dark grey lines, see Figures 12 and 13.

The grey values of each bubble correspond to the accuracy requirements taken from Table I. At this point it has to be recalled, that many of these

TABLE III
Performance summary. σ denotes commission errors only

Mission	Max. degree L	Resolution $\lambda/2$ (km)	σ_N @res. (cm)	σ_{Δ_g} @res. (mGal)
1. Very High Altitude Monitoring Missions				
1a	15	1333	100	1
1b	30	667	30	1
2. Gradiometry Missions				
2a	420	48	5	2
2b	500	40	3.5	2
2c	720	28	1.5	2
2d	900	22	1.5	2
2e	1200	0.15°	–	–
3. Low-Low SST Missions				
3a	150	133	10	2
3b	260	77	10	2
3c	220	91	10	3
3d	390	51	4	3
3e	–	–	–	–
3f	–	–	–	–
4. Configuration Flights				
4a	270	74	6	2
4b	–	–	–	–

accuracy requirements are based on rather rough estimates. The level line order is always from light grey at the left to dark grey at the right, corresponding to the increase of mission error curves to the short wavelengths. For all wavelengths to the left of a level line the accuracy is kept below the particular line level. The numbers corresponding to the dark grey lines, i.e. those with a signal-to-noise ratio of one, are also found in Table III (as σ_N@ res.).

Now, the bubbles' grey values can be compared to the level lines. Processes in bubbles with a certain grey tone to the left of the line in the same tone have a signal strength above the mission noise level and should therefore have a significant impact on the measured signal, whereas the other processes should become part of the noise. However, this comparison is a quite crude one, as the simulations are based on idealized concepts. The comparison does not consider the question whether single processes can be extracted from the total signal. Also the time resolution of mission scenarios with respect to gravity field variations has not been studied. The level lines in the figures end

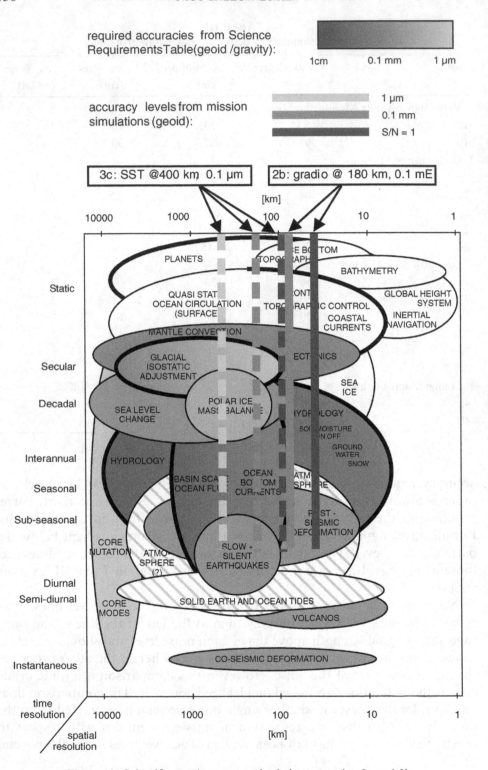

Figure 12. Scientific requirements and mission scenarios 3c and 2b.

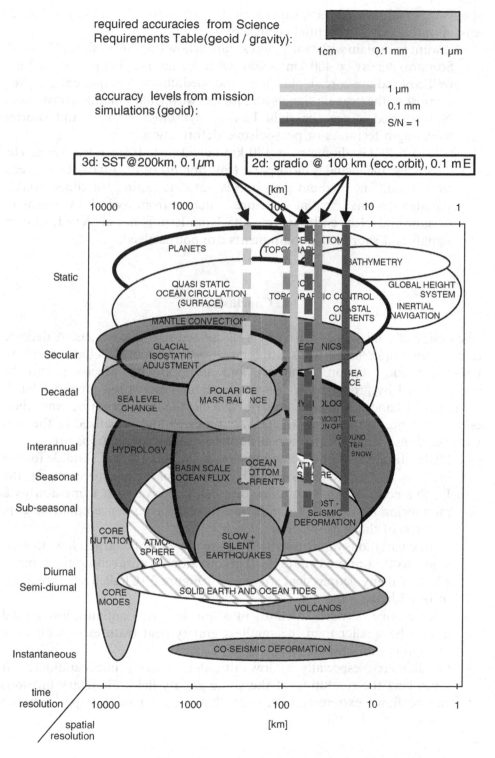

Figure 13. Scientific requirements and mission scenarios 3d and 2d.

at sub-seasonal time periods, supposing that all scenarios should produce at least monthly sets of potential coefficients.

We want to explain with two short examples how the figures should be read:

- Scenario 3c: SST @ 400 km height, with 0.1 μm ranging precision. Very well covered are core signals, long and medium GIA components, as well as most of the oceanic signals and longer wavelength mantle convection. Not covered is e.g. the right half of hydrology signals, and shorter wavelength tectonics or post-seismic deformation.
- Scenario 2b: Gradiometry @ 180 km height, with 0.1 mE precision. The dark grey line includes most parts (but not all) of the quasi-static ocean circulation. The medium grey line is very interesting for time-variable signals with small spatial structures, such as from hydrology, tectonics or slow and silent earthquakes. The 1 μm level is not achieved, so core signals and shorter GIA components are not resolved.

6. Conclusions

The design of a mission is driven by well-defined scientific targets. A delicate mixture of requirements of spatial resolution, temporal behaviour (both resolution and mission duration) and geoid/gravity precision can be accomplished by proper orbit design, choice of observation type and hardware performance. The outcome of specific mission scenarios and their correspondence to scientific targets was discussed and visualized in the previous sections. More general conclusions from these simulations are:

- High-altitude missions with a low–low SST observable are suitable for the investigation of long-wavelength phenomena, e.g. processes in the Earth's core. At high altitudes one can conceive such a mission as a monitoring mission. An application in planetary science could be core research of the giant planets.
- If time-variations of the gravity field are the scientific target, low–low SST seems to be a prime candidate. The question of unravelling the lumped effect of many geophysical sources, cf. Figure 3 (Challenge 1), remains to be addressed, though.
- The baseline between satellites in a low–low SST configuration should always be smaller than the smallest gravity field features to which the mission is still sensitive.
- Gradiometry, especially at low altitudes, is the prime candidate for improving the resolution of the static gravity field. Planetary missions can be flown extremely low, such that even a moderate-performance instrument will suffice.

- Formation flying is a promising concept and potentially a candidate for future gravity missions. The performance assessment in this study used certain simplifying assumptions, in particular the interpretation of the intersatellite links as a gradiometer. Future study into the characteristics of relative orbits and the modelling aspects of the corresponding observables is needed.
- High-eccentricity orbits with perigees going into the Earth's atmosphere could potentially lead to ultra-high resolution gravity missions. Such a mission would basically scan a certain latitude band. The technological aspects of such a demanding orbit were not addressed in this study. It was pointed out that with a proper perigee precession, the latitude bands of maximum resolution slowly precesses too, leading to a full scan of the Earth.

As mentioned earlier, the error simulations describe the performance of recovering a static gravity field. Time-variations must inherently be dealt with as a time-series of more-or-less static snap-shot solutions, e.g. monthly snapshots. Higher-frequency effects will necessarily alias into such monthly solutions, cf.(Han et al., 2004). This is a fundamental challenge to any future satellite mission, cf. Figure 3 (Challenge 2), which requires future research. For a more in-depth discussion of challenges 1 and 2, it is referred to Schrama, this issue).

References

Alfriend, K.T. and Schaub, H.: 2003, *AAS J. Astron. Sci.* **48**, 249–267.

Balmino, G., Perosanz, F., Rummel, R., Sneeuw, N., Sünkel, H. and Woodworth, P.: 1998, 'European Views on Dedicated Gravity Field Missions: GRACE and GOCE', An Earth Sciences Division Consultation Document, ESA, ESD-MAG-REP-CON-001, Noordwijk.

Bender, P. L., Hall, J. L., Ye, J. and Klipstein, W. M.:2003, in G. Beutler, R. Rummel, M. Drinkwater and R. von Steiger (eds.), *Earth Gravity Field from Space – from Sensors to Earth Sciences*, Space Science Series of ISSI, **18**, pp. 377–384, Kluwer Academic Publishers, Dordrecht.

European Space Agency: 1999, 'Gravity Field and Steady-State Ocean Circulation Mission', ESA SP-1233(1), report for mission selection of the four candidate earth explorer missions, ESA, Noordwijk.

Han, S. C., Jekeli, C. and Shum, C. K.: 2004, *J. Geophys. Res.* **109**(B04403), doi:10.1029/2003JB002501.

Jet Propulsion Laboratory: 1999, 'GRACE Science and Mission Requirements Document', JPL 327-200, Rev. B, JPL, Pasadena, CA.

McGuirk, J. M., Foster, G. T., Fixler, J. B., Snadden, M. J. and Kasevich, M. A.: 2002, *Phys. Rev. A* **65**, doi:10.1103/PhysRevA.65.033608.

Reigber, Ch., Schwintzer, P., Neumayer, K.-H., Barthelmes, F., König, R., Förste, Ch., Balmino, G., Biancale, R., Lemoine, J.-M., Loyer, S., Bruinsma, S., Perosanz, F. and Fayard, T.: 2003, *Adv. Space Res.* **31**(8), 1883–1888, doi:10.1016/S0273–1177(03)00162-5.

Rummel, R.: 2003, in G. Beutler, R. Rummel, M. Drinkwater and R. von Steiger (eds.), *Earth Gravity Field from Space – from Sensors to Earth Sciences* Space Science Series of ISSI, **18**, pp. 1–14, Kluwer Academic Publishers, Dordrecht.

Schrama, E. J. O.: 1991, *J. Geophys. Res.* **96**(B12), 20 041–20 051.

Schrama, E. J. O.: 'Impact of Limitations in Geophysical Background Models on Follow-on Gravity Missions', *this issue*.

Sneeuw, N.: 2000, 'A Semi-Analytical Approach to Gravity Field Analysis from Satellite Observations', Deutsche Geodätische Kommission, Reihe C, Heft Nr. **527**.

Earth, Moon, and Planets (2005) 94: 143–163
DOI 10.1007/s11038-004-5785-z

IMPACT OF LIMITATIONS IN GEOPHYSICAL BACKGROUND MODELS ON FOLLOW-ON GRAVITY MISSIONS

ERNST J. O. SCHRAMA

*Delft University of Technology, Faculty of Aerospace Engineering,
Astrodynamics and Satellite Systems, The Netherlands
(E-mail: e.j.o.schrama@lr.tudelft.nl)*

(Received 13 August 2004; Accepted 3 November 2004)

Abstract. Purpose of this article is to demonstrate the effect of background geophysical corrections on a follow-on gravity mission. We investigate the quality of two effects, tides and atmospheric pressure variations, which both act as a surface load on the lithosphere. In both cases direct gravitational attraction of the mass variations and the secondary potential caused by the deformation of the lithosphere are sensed by a gravity mission. In order to assess the current situation we have simulated GRACE range-rate errors which are caused by differences in present day tide and atmospheric pressure correction models. Both geophysical correction models are capable of generating range-rate errors up to 10 μm/s and affect the quality of the recovered temporal and static gravity fields. Unlike missions such as TOPEX/Poseidon where tides can be estimated with the altimeter, current gravity missions are only to some degree capable of resolving these (geo)physical limitations. One of the reasons is the use of high inclination low earth orbits without a repeating ground track strategy. The consequence is that we will face a contamination of the gravity solution, both in the static and the time variable part. In the conclusions of this paper we provide suggestions for improving this situation, in particular in view of follow-on gravity missions after GRACE and GOCE, which claim an improved capability of estimating temporal variations in the Earth's gravity field.

1. Introduction

Any gravity mission designed to map the temporal gravity field will inherently face the fact that oceanic tides and atmospheric pressure signals must be compensated for during the set-up of the normal equations containing the gravity parameters. Purpose of this article is to assess the consequence of this assumption, since the background corrections contain errors. In Section 2 it is explained that oceanic processes are not a primary objective of the GRACE mission, during the GRACE data processing all atmospheric pressure variations and oceanic mass variations due to tides are removed so that the continental hydrology signal remains as a primary signal to observe. In Section 3 we provide background information with respect to both background corrections. In Section 4 we will show that the accuracy of background models is insufficient to guarantee a full removal from the gravity solution (regardless whether it is static or temporal). A more rigorous approach is shown in Section 5 where we carry out a full simulation of both errors on the GRACE mission. In Section 6 we present our conclusions and recommendations.

2. The GRACE Mission

GRACE is designed to measure inter-satellite range-rates with an accuracy better than 10 μms^{-1} $Hz^{-1/2}$, for more details on the GRACE system and its ancestors (see Colombo, 1986; Dickey, 1997; Reigber et al., 2002) and the recent article by Tapley et al. (2004). The primary observation of the GRACE system is an integrated Doppler observation in the K band equivalent to that of a GPS carrier phase observation. This is equivalent to a biased range observation. In the near future the geodetic community will hopefully get access to data from another gravity mission, GOCE (cf. ESA, 1999), which is based on a different concept involving a gravity gradiometer. The goal of both missions is to map the Earth's gravity field whereby. GRACE will allow one to map the lower degree and orders of the field up to degree and order 120 while GOCE will be able to extend the resolution of this field to degree and order 250. The accuracy of the geoid obtained by both concepts depends on the length of the observation series that is used to create a gravity solution. GRACE has demonstrated a geoid accurate to about 3 mm which can be provided on a monthly basis according to Tapley et al. (2004). At the moment of writing GOCE is built by Astrium under contract from ESA, and the performance of this mission is assessed with analytical techniques as described in Schrama (1991).

Space born GPS receivers are a necessity for all gravity mapping missions. In the data processing scheme GPS tracking data is required to stabilize the least squares solution in the lower degrees. GPS information has to be weighted in some optimal way together with gradiometric or inter-satellite range-rate measurements. An example of a GPS-only tracking mission is CHAMP; in this case GPS is combined with accelerometry to map non-conservative forces caused by air drag, solar radiation pressure or other effects acting on the skin of the satellite. The CHAMP mission alone yields a significant improvement compared to earlier gravity solutions but is by far not capable of achieving the level attained by GRACE and GOCE where the performance is driven by the KBR instrument and the gradiometer respectively, (See also Reigber et al., 2002). During data reduction, i.e. all steps where normal equations of least squares systems are constructed, one will apply geophysical models to correct for known effects. The observation equations included in the normal matrices will be corrected for gravity, tidal effects, atmospheric pressure loading, measurement delay and offsets, and many other parameters. The differential improvement after inversion is then used to update our knowledge of parameters in the problem.

In the study presented here it is assumed that a follow-on gravity mission will be like the GRACE mission and that its focus will be on estimating temporal variations in the gravity field rather than the static field. For this purpose we will select two geophysical background models of which it is known that they are sufficiently large and where there are strong indications that the signal can not be fully represented by the model. Oceanic tides and atmospheric air pressure loading are suitable candidates for such models. The physics of both processes is well understood and it is likely that both will act as nuisance factors during data reduction. The actual scientific objectives of a follow-on gravity mission are very likely a study of the continental water balance (which is the largest signal) or variations of mass in the ocean interior or mass variations closer to the continental shelf margin, (See also Wahr et al., 1998).

Our starting point in the discussion is that ocean tide and atmospheric pressure models contain errors that propagate as systematic noise in the observations. Both processes have in common that they take place on time scales much shorter than the typical temporal resolution of the expected gravity field solution intervals (a month for GRACE). And for this reason it is expected that aliasing by background model errors will affect the performance of any follow-on gravity mission. If this assumption is true then the design criteria of a follow-on mission may need to be reviewed possibly in order to optimize or potentially benefit from a modified sampling strategy. Although the latter is perhaps desirable for the actual design of a future mission we will only provide suggestions in our conclusions.

3. Geophysical Effects

3.1. TIDES

Tides are the result of the gravitational attraction of Sun and Moon on the Earth itself, the relation to gravity missions is extensively described in Schrama (1995) where three tidal phenomena are identified.

Following the discussion in Schrama (1995), there is a direct tidal effect caused by the gravitational working of Moon or Sun directly on the satellite. This is the most straightforward part of the model and the accuracy of the correction depends on gravitational constants of the Earth and external bodies, relative position knowledge of these bodies, including position knowledge of the satellite. The relative accuracy of the direct tide effect is better than 10^{-8} so that direct tide model errors are not relevant for our study since we don't expect that planetary positions or their gravitational constants will be adjusted during data reduction.

In Schrama (1995) it is mentioned that there are two indirect tide effects which are caused by the deformation of the fluid and solid earth as a result of the direct tidal effect. A first indirect tide effect is the solid earth tide deformation model whereby we need to specify Love numbers k_n and h_n for $n = 2$ and 3 describing the elastic solid Earth response to external forcing. If horizontal site displacement must be adjusted then Love number l_n may need to be included in the model. Also for this part we expect no significant show stoppers albeit that a few Love numbers may need adjustment during data reduction of a new gravity mission.

The second indirect-tidal effect is related to ocean tide loading. This effect is far more difficult to model because a hydrodynamical model enters in the discussion. Ocean tide models contain many more parameters to specify the geographic response function which is now local rather than global. Significant progress has been made with the aid of the TOPEX/Poseidon altimetry (see also Schrama and Ray, 1994; Fu and Cazenave, 2001). As a result ocean tides are mapped to within 1.5 cm rms for the largest tidal constituent M_2 while the remaining constituents add less than 1.0 cm noise. The total rms. of the ocean tide signal is better than 3.0 cm rms in the deep oceans (see also Schrama and Ray, 1994). Yet the ocean tide model accuracy deteriorates on continental shelves and at latitudes beyond 66N or 66S because of the inclination of the TOPEX/Poseidon orbit. In Fu and Cazenave (2001) it is shown that M_2 errors in excess of 10 cm rms exist in coastal seas. In addition it is known that energy transfers from main tidal lines to parasitic ones as a result of non-linearity in the hydrodynamic equations in the quadratic bottom friction. Another reason is the presence of advective terms. Advection and bottom friction become relevant near the coast (see also Fu and Cazenave, 2001)

3.2. ATMOSPHERIC LOADING

Another correction that needs to be applied during data reduction deals with the weight of air masses that load on the Earth's surface. In earlier studies such as Velicogna et al. (2001) it is recognized that this effect is sufficiently large to be sensed by all gravity missions. The local weight of an air column is proportional to the terrain level pressure over continental areas. Over oceanic areas it is expected that the effect is compensated because of the inverted barometer (IB) mechanism. An algorithm more realistic than the IB model would include wind-stresses and air pressure forcing in a global hydrodynamic model. This method would modify the standard IB theory which after all assumes that there is an instantaneous -1 cm/mbar response of the sea level to air pressure variations. It is known that the global atmospheric pressure loading effect typically takes place on time scales of about 12 h and beyond and that the IB law becomes effective at time scales longer than three days (see also Mathers and Woodworth, 2001)

For all continental atmospheric loading calculations we have made an approximation for the weight of the air column loading the Earth's surface. The study of Verhagen (2001) has shown that radiosonde data can be approximated with an exponential decay law whereby we need as input the mean sea level pressure and the Earth's topography. For our purpose the largest uncertainty in the air pressure loading calculation comes from the accuracy of meteorologic models, and not so much the vertical distribution of mass in the air column which occurs between terrain level and the top of the atmosphere.

The largest input from the atmospheric loading effect on GRACE is expected over continental areas and not from incomplete compensation over oceanic areas. Meteorologic pressure models must be used during data reduction, well known meteorologic models are the NCAR reanalysis product and the ECMWF product. Both products are the result of a dynamic weather model in which meteorologic data as well as remote sensing data is assimilated. The accuracy by which the models differ is approximately 1.5 mbar, which is equivalent to a water layer of 15 mm (see also Velicogna, 2001). Atmospheric pressure loading errors typically occur on time scales less than the update interval of individual gravity solutions of a follow-on gravity mission. And therefore it is expected that some level of aliasing will take place as a result of the atmospheric pressure loading problem (see also Verhagen, 2001).

4. Degree Variance Signal and Error Spectra

Purpose of this section is to show degree variance spectra for air pressure variations and ocean tide variations and to convolve these input mass fields towards temporal changes in the geoid.

4.1. TIDES

To compute the degree variances of the ocean tide fields convoluted towards the geoid under the assumption of a self attraction formulation that includes lithospheric deformation we assume that tides are prescribed by (See also Cartwright, 1993).

$$\zeta = \sum_v H^v \cos(X^v - G^v), \tag{1}$$

where the in-phase and quadrature components of each wave with index v are:

$$P^v = H^v \cos(G^v),\tag{2}$$

$$Q^v = H^v \sin(G^v).\tag{3}$$

For each constituent v the maps P^v and Q^v are approximated in spherical harmonics (now dropping index v):

$$P = \sum_{nma} A_{nma} Y_{nma}(\theta, \lambda),\tag{4}$$

$$Q = \sum_{nma} B_{nma} Y_{nma}(\theta, \lambda),\tag{5}$$

where the dimension of P and Q and hence A and B is meters. The corresponding convolution towards geoid heights yields the spherical harmonic coefficients C and D (cf. Schrama, 1997)

$$\left\{ \begin{matrix} C_{nma} \\ D_{nma} \end{matrix} \right\} = g^{-1} \frac{3\mu(\rho_w/\rho_e)}{a_e^2(2n+1)} (1 + k'_n) \left\{ \begin{matrix} A_{nma} \\ B_{nma} \end{matrix} \right\},\tag{6}$$

where g is the gravitational acceleration, μ the gravitational constant, ρ_w and ρ_e are the density of sea water and the mean density of the Earth, k'_n are load Love numbers, and a_e is the mean equatorial radius. The degree variances for the geoid are

$$E_n^2 = \frac{1}{(2n+1)} \sum_{ma} [C_{nma}^2 + D_{nma}^2].\tag{7}$$

For the simulation of tide model errors we difference the coefficients C and D from two ocean tide models to obtain δC and δD. The simulated tide model error degree variance δE_n^2 is then

$$\delta E_n^2 = \frac{1}{(2n+1)} \sum_{ma} [\delta C_{nma}^2 + \delta D_{nma}^2].\tag{8}$$

4.2. ATMOSPHERIC PRESSURE VARIATONS

In order to compute the average degree variance of a sequence of equivalent water height fields that follow from an IB model including an error assessment of this average we proceed as follows. Water level variations as a result of air pressure variances are simplified by (see Gill, 1982)

$$\zeta = \frac{-1}{g\rho}(P - P_0),\tag{9}$$

where P_0 is some reference pressure value. Here P_0 is 1013.3 h Pa while g is the gravity acceleration (9.81 m/s^2) and $\rho = 1026$ kg/m^3. P follows from a meteorologic model, for which we have used the ECMWF and the NCAR reanalysis model which come as daily grids during 1992. In this case the values of ζ only exist on land, and the air pressure difference term $(P - P_0)$ is scaled down by an exponential law from the sea level to the terrain level (see Verhagen, 2001). Over sea the ζ values are set to zero and full mass compensation is assumed in agreement with the inverse barometer law.

The convolution of ζ (now provided as a spherical harmonic coefficient set in terms of coefficients A_{nma} at time step i) to geoid heights is similar to that of tides (see Schrama, 1997)

$$C_{nma,i} = g^{-1}\frac{3\mu(\rho_w/\rho_e)}{a_e^2(2n+1)}(1 + k'_n)A_{nma,i},\tag{10}$$

whereby geoid heights are calculated on the Earth's surface. The degree variances for the geoid are now computed as

$$E_{ni}^2 = \frac{1}{(2n+1)}\sum_{ma}C_{nma,i}^2.\tag{11}$$

At this point we define the average degree variance of a sequence of I_{max} pressure grids as follows

$$E_n = \frac{1}{I_{max}}\sum_{i=1}^{I_{max}}E_{ni}.\tag{12}$$

For the simulation of model errors (and then the self attraction and loading representation) we difference the coefficients A in Equation (10) from two models to replace them by δA.

4.3. RESULTS

The computed degree variances for the tide calculation involves the models FES99 and GOT99.2. We are aware of the fact that newer versions of such models will continue to evolve from the tidal community. For the scope of this study we do not expect that the results significally change because the set of observed tidal constants is heavily biased towards the same TOPEX/ Poseidon observations. For the meteorologic models we have used ECMWF and NCAR reanalyis daily pressure grids in 1992. Also here we are aware of the fact that 1992 is taken as a reference year and that more modern versions exist. For a simulation study such as this we have no indication that our conclusions are significantly affected. Square roots of degree variances of all relevant data are shown in Figure 1. The conclusion of this calculation is that below degree and order 50 the degree variance errors of both tides and air pressures are larger than the initially advertised performance noise of GRACE, which is about 0.01 mm at the lowest degrees according to the prelaunch estimates (see also Dickey, 1997). The same problem will also play a role with any follow-on gravity mission declicated to the observation of temporal gravity. Tide and atmosphere errors are at the moment of writing a limiting factor up to degree and order 10–15, a region in the gravity field where many geophysical signals leave their signature. Discussion on basis of degree variance spectra:

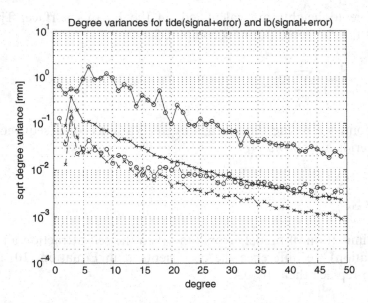

Figure 1. Square root of degree variances of tides and atmospheric pressure loading and simulated errors, horizontal axis degrees, vertical axis: meter geoid change. The solid line with circles represents tide signal and the dashed line with circles are tide errors. The solid lines with crosses represents atmospheric pressure, and the dashed line with crosses follow from the simulated atmospheric error.

- Usually the T/P tide model accuracy specifications are based on deep water comparisons to tide gauges where nowadays a 3 cm rms total error is found. Over continental shelves localized errors still exist and the models are some-times up to 15 cm or higher in error. In polar regions there is no T/P coverage and the realism of the model depends on hydrodynamic models such as FES99. Overall tide model errors are greater than air pressure errors as far as temporal changes in the geoid are concerned (see also Figure 1).

- There exists a coupling mechanism between tides and air pressure in the form of atmospheric tides. This effect is the consequence of the atmosphere being forced as an air mass layer that experiences gravitation forcing in the same way as the oceans. Aliasing of S_1/S_2 oceanic and air tides is relevant for sun-synchronous orbits which are considered for GOCE (but not for CHAMP and GRACE). This effect will result in a pseudo static field mapping along on the ground tracks of GOCE. Errors in S_1 or S_2 models, either oceanic or atmospheric, will therefore alias into a static gravity field error (see also Schrama, 1995).

- For the non-tidal air pressure signal we know that the in-situ point wise now-cast error for calm or normal weather condition is approximately 1–1.5 mbar. This value is typical for both ECMWF and NCAR reanalysis data (Velicogna, 2001) Averaging over space and time helps to drive down this error, but levels better than 0.3 mbar are unlikely at the moment of writing according to Velicogna (2001). Yet meteorologic models have heterogeneous error characteristics and we know that some regions are more poorly represented than others. Antarctica is a typical region where NCEP reanalysis data and ECMWF data significantly differ, ie. the errors will be larger. We ignored these effects in the computation of our degree variances by cutting out all latitudes pole wards of 70N and 70S.

- It was found in separate studies that meteorologic models use their own topography (or orography) which is adapted to the numerical scheme for solving the differential equations. This effect becomes visible as a difference between meteorologic models and is correlated with topographic height.

- More serious is the conclusion that meteorologic pressure grids are provided on a 3 hourly to daily basis while the temporal gravity solution interval is typically 10 days to a month (for GRACE 6 hourly fields are used). This means that errors in the atmospheric pressure variations will alias into the gravity solution.

- Finally it should be remarked that oceanic responses are modelled such that they behave like an inverted barometer, i.e. masses are fully compensated over the ocean and in reality we know that this is not the case, see for instance (Mathers and Woodworth, 2001). The IB model may contain errors up to 20% for periods up to a day and there exist resonance regions in the Southern hemisphere that cause systematic errors of a few percent. The GRACE team relies on an alternative IB barotropic model forced by air pressure and wind that is used during data reduction.

- From the degree variance spectra it is clear that the noise level of the projected gravity fields (i.e. projected with analytical propagation techniques as discussed in Dickey (1997), ESA (1999) and Schrama (1991)) will be too optimistic. At least one reason is the existence of model noise from two main geophysical corrections which must be applied during data reduction of either GRACE or a follow-on gravity mission. The analytical propagation techniques on which the results in Dickey (1997) and ESA (1999) are biased in this sense and do not include these effects. Hence we conclude that these analytical gravity mission performance curves are probably too optimistic in the lower degrees. This may also partially explain why all GRACE hydrology results as shown in Tapley et al. (2004) avoid the use of degree 2 spherical harmonics and the fact that Gaussian smoothing with a 400 km averaging radius is required to suppress noise in their gravity solutions.

5. GRACE Simulation Experiment

In order to simulate the inter-satellite range-rate signal as a result of geophysical model errors we use the existing GRACE gravity mission trajectory along which suitable potential functions are simulated. The following sections describe the choice of the baseline orbit and the simulation experiment.

5.1. CHOICE OF THE GRAVITY MISSION BASELINE ORBIT

The simulations depend on the choice of a reference trajectory for which we have chosen the nominal GRACE trajectory. Initial orbital elements have been selected from the GRACE web site at the university of Texas at Austin, Center of Space Research: $a = 6861124.723$ m, $e = 0.001687$, $I = 89.001°$, $\Omega = 307.659°$, $\omega = 17.338°$ and $f = 307.052°$ on 7/3/2003 14:37:00. For the gravity model we have used the EGM-96 model (see Lemoine et al; 1998), complete to degree and order 70. Furthermore direct astronomical and indirect solid Earth tides have been modelled where JPL's DE200/LE200 model provides planetary and lunar locations; in addition relativistic effects are also taken into account while IERS bulletin B values are used for the definition of pole positions.

Although in reality the GRACE satellites will go through various orbit regimes this simulation relies on a baseline trajectory which gives a reasonable track coverage in about 9 days. In fact, the simulated ground tracks repeat within 1 degree longitude variation in the ascending node every 8.9747 days when the GRACE system completes 137 orbits. To simplify the discussion it is assumed that this ground track pattern is repeated every 137 orbital periods. The obtained sampling characteristics are used in the tidal aliasing experiment.

5.2. SIMULATION MODEL FOR TIDES

In this section we will discuss a quick diagnostic model to explain inter-satellite variations on the GRACE system as a result of geophysical model noise. In all computations we will assume that both systems are separated by about 30 s so that the inter-satellite separation distance is about 230 km. To simulate this inter-satellite range-rate effect we assume for simplicity: (1) the total energy being the sum of potential and kinetic energy is conserved for GRACE 1 and 2, i.e. the non-conservative forces acting on GRACE 1 and 2 are ignored,(2) both GRACE satellites follow the same trajectory and are only separated in time,(3) the along track velocity component is differenced between both satellites to simulate inter-satellite range-rate variations,(4) both satellites move with an average velocity v_0, (5) there are no coupling terms to earth rotation. Under these assumptions the inter-satellite velocity variations between GRACE 1 and 2 are equal to the difference of the simulated error in the potential ΔU's of each satellite

$$\Delta v(t_1, t_2) = v_0^{-1}(\Delta U(t_2) - \Delta U(t_1)), \tag{13}$$

where v_0 is local velocity at the reference orbit. For the ocean tide model we have the following relation:

$$U(r, \phi, \lambda, t) = \sum_i A_i(r, \phi, \lambda) f_i \cos(\chi_i + u_i) + B_i(r, \phi, \lambda) f_i \sin(\chi_i + u_i), \tag{14}$$

where i is a running index over tidal waves in the model, χ_i f_i and u_i are astronomically defined quantities related to the definition of these waves (see Cartwright, 1993), while A_i and B_i denote in-phase and quadrature terms which are defined as follows

$$A_i(r, \phi, \lambda) = \sum_{nma} \frac{3\mu(\rho_w/\rho_e)}{a_e^2(2n+1)}(1 + k'_n)\left(\frac{a_e}{r}\right)^{n+1} A_{nma,i} Y_{nma}(\phi, \lambda), \tag{15}$$

$$B_i(r, \phi, \lambda) = \sum_{nma} \frac{3\mu(\rho_w/\rho_e)}{a_e^2(2n+1)}(1 + k'_n)\left(\frac{a_e}{r}\right)^{n+1} B_{nma,i} Y_{nma}(\phi, \lambda). \tag{16}$$

In these expressions μ is the Earth's gravitational constant, ρ_w is the mean density of sea water, ρ_e is the mean density of rock, a_e is the Earth's equatorial radius, k_n' are elastic load Love numbers, Y_{nma} are normalized spherical harmonic functions, index a selects the combination $\cos m\lambda \bar{P}_{nm}(\sin \phi)$ or $\sin m\lambda \bar{P}_{nm}(\sin \phi)$ with $\bar{P}_{nm}$ representing normalized associated Legendre functions (see also Ray et al., 2003). The terms $A_{nma,i}$ and $B_{nma,i}$ follow directly from a spherical harmonic analysis of ocean tide model maps. To simulate tide model errors we have used the FES99 and the GOT99.2 models (see Ray et al., 1999; Lefèvre, 2002) to construct the corresponding spherical harmonic coefficients. It obtain ΔU that represents tide model errors we differenced $U(\text{FES99})$ and $U(\text{GOT99.2})$.

5.3. SIMULATION RESULTS

Equation (13) in combination with the definition of the ocean tide error potential gives a series of Δv values along the baseline orbit. In fact, Δv is now easily projected ahead in time because it is a harmonic function; whereby we assume a repeat cycle length of 8.9747 days. For this purpose our simulation program works in two steps. We start by computing the harmonic coefficients for all selected tidal waves (only 8 are used) and we implement these harmonic coefficients in a time series generation algorithm provided by R.D. Ray.

A first result is to find extreme excursions over each $1° \times 1°$ block over the sphere; this is shown in Figure 2. From this Figure we observe that most parts of the globe experience velocity errors less than 1 μm/s. In quiet coastal zones we see that the velocity variations are of the order of 1–2 μm/s and in certain noisy regions we see very localized errors of 10 μm/s or more. Such phenomena happen over certain continental shelves where it is known that the global ocean tide models are inaccurate. A moderate smearing of this phenomenon takes place because of the upward continuation from the Earth's surface to the potential function at satellite height. Similar large excursions are observed at latitudes beyond 66N and 66S which are in our opinion due to the quality of tide models beyond the TOPEX/Poseidon inclination latitude.

5.4. STRATEGY FOR ERROR SUPPRESSION

It is evident from the result in Figure 2 that GRACE velocity data can easily contain errors caused by tidal modelling that exceed the measurement accuracy. The question is now whether one can accept such errors or whether additional nuisance parameters need to be defined. The answer to this

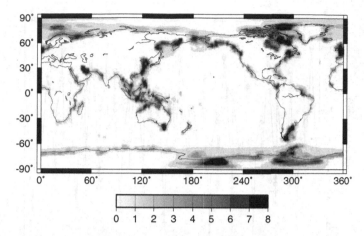

Figure 2. Extreme velocity variation observed in the simulation set where velocity errors are projected using the energy conservation approach from simulated ocean tide model errors. Scale: μm/s.

question is probably that it is desirable to estimate suitable nuisance parameters, i.e. it is realistic to assume that future activities will concentrate on the estimation of unmodelled tidal effects. But we will also warn that modelling tide errors in GRACE is far from trivial due to unfavorable sampling of short periodic tides compared to the, relative long repeat cycle of GRACE. (In reality GRACE doesn't have a repeat cycle, the longitudes of the nodes of the ascending ground tracks coincide to within $1°$ in a 8.9747 day mapping cycle)

Figure 3 shows for instance the history of all collected velocity residuals in a radius of $1°$ around bin 65N 80W, which is in the Hudson bay. If the Δv signal were favorably mapped then we would easily recognize periodic features in this series, instead we get to see that the velocity errors appear in local clusters that seem to alternate every 9 days in sign. In fact, in order to be able to recognize an aliased beat signal due to unmodelled tides, it is necessary to extend this experiment over many more repeat cycles.

More evidence for this observation is provided in Figure 4 where an attempt is made to recover by means of a least squares filter tidal amplitudes and phases for each $1° \times 1°$ bin where 200 repeat cycles are used. After correcting the GRACE data for all solved for corrections per bin we are able to present the velocity excursions in the same way as in Figure 2. But even after this filtering step it is obvious that we are not able to suppress all tidal errors, i.e. significant residuals remain visible in Figure 4 which is a likely indicator that it will be difficult to undo the GRACE data set from tidal modelling errors. Similar results are presented in Knudsen and Andersen (2002) and Ray et al. (2003).

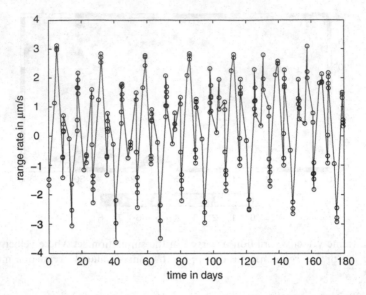

Figure 3. History of velocity variations observed over the Hudson bay 65N 80W, along the *x*-axis the time is shown in days, along the *y*-axis the inter-satellite velocity variations are shown in μm/s.

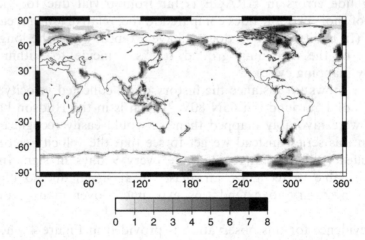

Figure 4. Extreme velocity variations observed in the simulation set after we took care of removing a 16 parameter function for each $1° \times 1°$ bin. Scale: μm/s.

5.5. HOW REALISTIC IS THE ENERGY CONSERVATION APPROACH

As was pointed out, the here described energy conservation approach with statistics displayed in Figure 2 is an approximation. A better method is to simulate range-rate errors by means of the GEODYN orbit computation program whereby a 10 day trajectory was integrated once using the GOT992 ocean tide model. During data reduction, whereby initial state vectors of both GRACE's are solved for, it is assumed that the inter-satellite range-rate and position) knowledge of the two orbiters are observations. In this second (iterative) orbit adjustment process the FES99 model is used as a forcing model. No effort was undertaken to model skin accelerations on both GRACE satellites during this run. The inter-satellite range- rate observations are binned in $1° \times 1°$ blocks in the same manner as Figure 2.

In Figure 5 we observe an abundance of low frequency variations in the Δv's which don't seem to correspond to the earlier results obtained with the energy conservation approach where we found localized velocity excursions around geographic regions where coastal tide model errors occur.

An obvious explanation of this phenomenon follows from the orbit dynamics experienced by two satellites that translates itself in an inter-satellite range-rate at once and twice per orbital period. A GEODYN simulation will also reveal that the leading GRACE satellite experiences velocities and accelerations along a slightly different flight path compared to the trailing GRACE satellite. Furthermore in this simulation the velocity vector of each satellite doesn't project directly on the inter-satellite range. Our simplified energy conservation approach does not include all complexities which are part of the GEODYN simulation and ignores the long wavelength effect visible in Figure 5.

In an attempt to suppress long wavelength contamination we implemented a high pass filter with a cut-off frequency at 3000 s. After implementation of this filter the Δv's show a behavior as in Figure 6. These results show similar characteristics as obtained with our energy conservation approach, i.e. the Δv effect is increased over regions where there are large tidal modelling errors. Nevertheless there remain some remarkable differences which we blame for

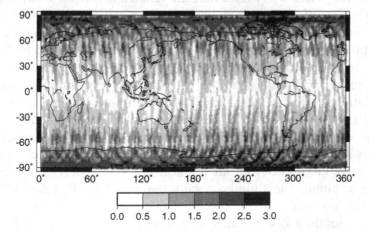

Figure 5. Extreme velocity variations observed in the GEODYN simulation set. Scale: $\mu m/s$.

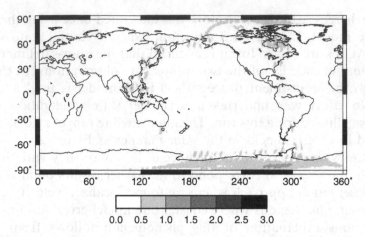

Figure 6. Extreme velocity variations observed in the GEODYN simulation set after application of a 3000 s high pass filter.

the moment to the realism put into the GEODYN simulation where the simulation runs over 10 days and ocean tide spherical harmonic coefficients to represent FES99 and GOT992 are truncated at degree and order 35. These limitations suppress localized excursions of the tides in shallow seas and avoid that the ocean tide signal is mapped over more cycles.

5.6. SIMULATION MODEL FOR AIR PRESSURE VARIATIONS

In order to study air pressure variations it is necessary that we evaluate the effect of air pressure changes and its direct consequence on a follow-on gravity mission. Here we remind that the realism of such a study depends on the specification of meteorologic model errors which are on the one hand difficult to quantify largely because similar techniques and observation data are used by different meteorologic centers. On the other hand we know that there are geographical regions such as the Antarctic where both pressure fields are substantially different and where one or both models have significant limitations (see also Verhagen, 2001). Estimates for meteorologic pressure errors can be found in Velicogna (2001), and a more complete simulation of the effect on GRACE including a full formal assessment of such errors in the recovery of science signals from GRACE such as discussed by Wahr et al. (1998) does in our opinion not yet exist. The approach used here is to evaluate the difference between the NCAR reanalysis surface pressure models and the ECMWF surface pressure model whereby the atmospheric loading is contained in continental areas and where an exponential decay law, for detail (see Velicogna, 2001) is used to convert sea level pressures into terrain level pressures. Mass variations as a result of the simulated air pressure differences are then converted into equivalent water height. It is the gravitational effect of this layer and the application of relations similar to Equations (15) and (16) that yield a potential ΔU to be substituted in Equation (13).

Figure 7 shows extreme velocity variations observed over one by one degree bins on the globe as they are encountered in a GRACE simulation set with a length of 1 year. From this Figure we conclude that the differences between the models reach 3 μm/s; such errors will be significant for GRACE or any follow on gravity mission. Moreover we observe that the Δv error pattern is geographically constrained to Asia, North America, and the Antarctic. In our calculations we ignored meteorologic pressure difference at latitudes beyond 70N and 70S due to unrealistically large meteorologic modelling errors as discussed in Verhagen (2001).

Any effort to reduce air pressure errors will require to design filters that are even more complicated than to reduce periodic modelling errors such as for tides. The design of such filters could exploit geographical or temporal properties of the signal. Geographical interpretation: Evidently, Figure 7 shows that meteorologic errors are not only constrained to the coastal zones as is the case with ocean tide errors, but rather that the error appears to be correlated to land topography. Furthermore it is evident that variations in the tropics are less than those at higher latitudes. Temporal interpretation: Air pressure variations and their errors do contain daily and seasonal signals but lack for a significant part astronomic periodicity. At best averaging procedures will therefore help to suppress the errors (see also Velicogna et al., 2001).

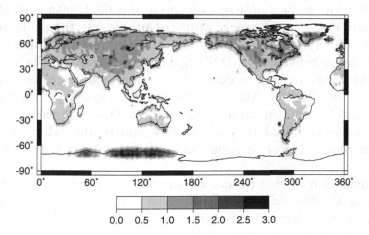

Figure 7. Extreme velocity variations observed in the simulation set as a result of the air pressure error simulated as the difference between ECMWF and NCAR reanalysis data. Scale: μm/s.

6. Conclusions and Recommendations

This article started with the scientific rationale of the GRACE mission, and the remark that certain geophysical corrections may pose a limitation for the interpretation of GRACE observation data. This conclusion follows directly from the degree variance spectra of signal and noise (see Figure 1) when they are overlaid on the gravity mission performance spectra such as shown in Dickey (1997) and ESA (1999). Both studies did not account for aliasing effects as a result of geophysical model errors.

During the GRACE data reduction a self attraction potential for air pressure and ocean tides will be computed during the procedure, ie. the gradients of Equation (14) or equivalent will be used in the orbit computation scheme. Purpose of this study is to focus on the consequences of modelling errors in such potentials on the inter-satellite range-rate variation between both GRACE satellites. Our study is diagnostic and merely intends to identify and characterize the nature of the inter-satellite range-rate errors. Furthermore we hope to draw conclusions from this study in order to assess the consequences for any possible follow-on gravity missions.

6.1. CONCLUSION FROM THE GRACE SIMULATION EXPERIMENT

For tides we observe that velocity errors of the order of 1–10 μm/s occur which are caused by model differences in continental shelf areas and polar areas. One reason is the spatial resolution of the T/P altimeter data that was used in the construction of GOT99.2 and FES99 and the tendency of tidal surface waves to spatially narrow which depends on coastal geometry and bathymetry. The surface propagation speed of a tidal wave is equivalent to $\sqrt{gH}$, and for seas or channels with a depth H less than 5 m the depth integrated velocity times tidal period becomes less than the inter-track separation distance between two TOPEX/Poseidon ground tracks at the equator, i.e. there may be surface details in the tides which are too small to be observed by the altimeter. Another reason for tides to differ in these regions is that the models only contain astronomically defined frequencies while they miss overtones or mixed frequencies as would be the case when realistic dissipations due to bottom friction and advection were part of a dynamic tide model.

The simulation of air pressure errors and the upward continuation to a potential at satellite height was performed with the aid to ECMWF and NCAR reanalysis data. In this approach we have used the energy conservation approach and have simulated this effect over a 1 year period. For this effect we observe that the Δv error pattern leads to excursions of approximately 3 μm/s and that the error pattern is geographically constrained to

Asia, North America, and the Antarctic. In addition we see that there is a tendency for correlation of such errors with the local topography. No further attempts have been made to reduce this type of error with the aid of specifically designed filters for this problem.

6.2. THE NEED FOR MITIGATION STRATEGIES

So far we conclude that tide errors for the main constituents could be estimated provided that gravity field mapping missions are designed such that the corresponding frequencies do not alias to infinite periods. Atmospheric tides like S_1 and S_2 are capable of adding a permanent contribution to the static gravity field. An example of a simulated tide error can be found in Schrama (2003), is shown for the S_2 tide on page 188, it is typical for a static contribution to the gravity field. To mitigate tide model errors it is necessary to optimize the baseline orbits for follow-on gravity missions. Another option is to include suitable nuisance parameters in data reduction scheme.

We can expect that tide and atmospheric pressure models will probably improve in the next decade. For atmospheric models we will probably see that GPS limb sounding could help to independently and globally map the atmosphere and its density variances.

Finally we want to remark that scientific disciplines may develop their own feature extraction techniques to estimate their signals. Hydrology studies could focus on optimized anti-leakage techniques that drive the errors down outside selected river basins, moreover they could benefit from the presence of annual periodicity in the hydrologic cycle.

From the results so far shown by the GRACE science team we can expect that the annual hydrology signal in the geoid is larger than the geoid error to expect from this study. A propagation of tide model difference between FES99 and GOT99.2 results in geoid features no larger than about 5 mm. In Tapley et al. (2004) and during presentations of the GRACE science team at the EGU in Nice in 2004 it was demonstrated that the hydrology signals sensed by GRACE results in geoid signal of approximately 10 mm with a clear annual period. Spatial averaging to suppress trackiness in the GRACE geoid solutions is required to obtain the hydrologic signal (see Tapley et al., 2004; Wahr, 2004).

6.3. FUTURE WORK

From the GRACE simulation we conclude that the energy conservation approach and the GEODYN approach resulted in different answers which are probably due to the realism put in the first approach. We intend to increase the level of realism that is put in our current GEODYN simulation, essentially by increasing the spherical harmonic expansions of the used ocean tide field and by increasing the length of the simulation data set. Similar plans need to be worked out for the air pressure error simulation where so far we have only relied on the energy conservation approach.

The identified error patterns of both geophysical effects are however significant and large enough to affect temporal solutions of the gravity field by GRACE, especially when one would attempt to recognize smaller signals. The consequences of geophysical background model errors and their effect on a gravity inversion is however not part of this study but is in progress (Visser and Schrama, 2004).

Another recommendation is to investigate alternative strategies that exploit the synergy of different gravity missions possibly in combination with auxiliary measurements.

Acknowledgements

This study was funded by Astrium under an Enabling Technology study contract. Pieter Visser provided the results from the GEODYN simulation, the NCAR reanalysis data was provided by Cathy Smith at NOAA and the ECMWF pressure data became available to DEOS in a cooperation with GRGS who obtained the data from Meteo-France. The author thanks Bert Vermeersen and an unknown reviewer for suggestions that have improved this manuscript.

References

Cartwright D. E.: 1993, 'Theory of Ocean Tides with application to Altimetry, Lecture Notes in Earth Sciences', *In* Reiner Rummel and Fernando Sansò (eds.), *Satellite Altimetry in Geodesy and Oceanography*, Springer Verlag, Berlin, Vol 50.

Colombo, O. L: 1986, 'The global mapping of gravity with two satellites', Netherlands Geodetic Commission, New Series No. 30, pp. 1–180.

Dickey, J.O.: 1997, 'Satellite Gravity and the Geosphere', National Research Council Report. National Academy Washington D.C. 112 pp.

ESA.: 1999, 'Gravity field and steady state ocean circulation mission', ESA SP-1233, pp 217.

Fu L.L and Cazenave A.: 2001, Satellite altimetry and Earth Sciences, A handbook of techniques and applications', International Geophysics Series, Academic press, New York, Vol 69.

Gill A. E.: 1982, *Atmosphere – Ocean Dynamics*, Academic Press, New York, Vol 30.

Knudsen, P. and Andersen, O.: 2002, 'Correcting GRACE gravity fields for ocean tide effects' Geophysical Research Letters **29**(8), 19.1–19.4.

F. Lefèvre, Lyard, F., Le Provost, C., and Schrama, E.: 2002, 'FES99: A global tide finite element solution assimilating tide gauge and altimetric information', *Journal of Atmospheric and Oceanic Technology* **19**, 1345–1356.

Lemoine et al.: 1998, The development of the joint NASA GSFC and NIMA geopotential model EGM-96 gravity model, NASA/TP-1998-206861.

Mathers, E.L. and Woodworth, P. L.: 2001, 'Departures from the local inverse barometer model observed in altimeter and tide gauge data and in a global barotropic numerical model', Journal of Geophysical Research Vol. 106, No. C4, p. 6957 (2000JC000241).

Ray, R.: 1999, 'A Global Ocean Tidel Model from TOPEX/POSEIDON Altimetry: GOT99.2, NASA TM 1999–209478.

Ray R., Rowlands, D., and Egbert, G.: 2003, 'Tidal models in a New Era of Satellite Gravimetry', *Space Science Reviews* **108**, 271–282.

Reigber, Ch., Balmino, G., Schwintzer, P., Biancale, R., Bode, A., Lemoine, J-M, König, R., Loyer, S. Neumayer, H., Marty, J-C., Barthelmes, F., Perosanz, F., and Zhu, S.Y.: 2002, 'A High-Quality Global Gravity Field Model from CHAMP GPS Tracking Data and Accelerometry (EIGEN-1S), Geophysical Research Letters, Vol.29, No. 14, 10129/2002GL015064.

Schrama, E. J. O.: 1991, 'Gravity Field Error Analysis: Applications of Global Positioning System Receivers and Gradiometers on Low Orbiting Platforms', *Journal of Geophysical Research* **96**(B12), 20, 041–20,051.

Schrama, E. and Ray, R.: 1994 'A preliminary tidal analysis of TOPEX/POSEIDON altimetry', JGR Vol 99, No C12, pp.24, 799–24,808.

Schrama, E.: 1995, 'Gravity missions reviewed in the light of the indirect ocean tide potential. in proc. of *IAG symposia proceedings* No 116, pp 131–140, Springer Verlag, Berlin.

Schrama, E.: 1997, 'Satellite Altimetry, Ocean Dynamics and the Marine Geoid, Lecture Notes in Earth Sciences', in Reiner Rummel and Fernando Sansò (eds.), *Geodetic Boundary Value Problems in View of the One Centimeter Geoid*, Springer Verlag, Berlin, Vol 65.

Schrama, E.: 2003, Error 'Characteristics Estimated from CHAMP, GRACE and GOCE derived geoids and from satellite altimetry derived mean dynamic topography', *Space Sciences Reviews* **108**, 179–193.

Tapley, B. D., Bettadpur, S., Ries, J. C., Thompson, P. F., and Watkins, M. M.: 2004, 'GRACE Measurements of Mass Variability in the Earth System', *Science* **305**, 503–505.

Velicogna, I., Wahr J., and Vandel Dool, H.: 2001 'Can surface pressure be used to remove atmospheric contributions from GRACE data with sufficient accuracy to recover hydrological signal?', *Journal of Geophysical Research* **106**(B8), 16415.

Verhagen, S., Time variations in the gravity field, the effect of the atmosphere, MSc report, 81 pages, Delft University of Technology, Department of Geodesy, 2001.

Visser P. N. A. M., and Schrama E. J. O. Space-borne gravimetry: how to decouple the different gravity field constituents, Submitted to IAG GGSM04 conference proceedings, 2004.

Wahr, J., Molenaar, M., and Bryan, F.: 1998, 'Time variability of the Earth's gravity field: Hydrological and oceanic effects and their possible detection using GRACE', *Journal of Geophysical Research* **103**(B12), 30205–30229.

Wahr, J.: 2004, Personal communication and presentations at EGU meeting Nice, France.